INDISTRACTABLE

INDISTRACTABLE

How to Control Your Attention
and Choose Your Life

NIR EYAL

WITH JULIE LI

BLOOMSBURY PUBLISHING
LONDON · OXFORD · NEW YORK · NEW DELHI · SYDNEY

BLOOMSBURY PUBLISHING
Bloomsbury Publishing Plc
50 Bedford Square, London, WC1B 3DP, UK

BLOOMSBURY, BLOOMSBURY PUBLISHING and the Diana logo are trademarks of
Bloomsbury Publishing Plc

First published in Great Britain 2019

Bookmark by Muneer A. Safiah from the Noun Project

The identities of some people in this book have been changed. Some sections have
previously appeared on the author's blog, NirAndFar.com

For legal purposes the Acknowledgements on p. 279
constitute an extension of this copyright page

Bloomsbury Publishing Plc does not have any control over, or responsibility for, any
third-party websites referred to in this book. All internet addresses given in this book were
correct at the time of going to press. The author and publisher regret any inconvenience
caused if addresses have changed or sites have ceased to exist, but can accept no
responsibility for any such changes

A catalogue record for this book is available from the British Library

ISBN: HB: 978-1-5266-1022-5; TPB: 978-1-5266-1021-8; EBOOK: 978-1-5266-1023-2

2 4 6 8 10 9 7 5 3 1

Typeset by Newgen KnowledgeWorks Pvt. Ltd., Chennai, India
Printed and bound in Great Britain by CPI Group (UK) Ltd, Croydon CR0 4YY

To find out more about our authors and books visit www.bloomsbury.com
and sign up for our newsletters

For Jasmine

CONTENTS

AN IMPORTANT NOTE

Before you start reading, make sure to download the supplementary materials from my website. You'll find free resources and downloads, as well as my latest updates, at:

NirAndFar.com/Indistractable

Most importantly, you'll want to use the accompanying workbook, which I designed with exercises for each chapter to help you apply what you learn to your own life.

Also, please note that I do not have a financial interest in any of the companies mentioned unless specifically stated and my recommendations are not influenced by any advertisers.

And, if you'd like to get in touch personally, you can reach me through my blog at NirAndFar.com/Contact.

Introduction: From *Hooked* to *Indistractable*

There's another yellow book you'll find on the shelves of most major tech companies. I've seen it at Facebook, Google, PayPal and Slack. It's given out at tech conferences and company training events. A friend working at Microsoft told me the CEO, Satya Nadella, held up a copy and recommended it to all of the company's employees.

The book, *Hooked: How to Build Habit-Forming Products*, was a *Wall Street Journal* bestseller and, at the time of writing, still ranks as the number one book in the 'Products' category on Amazon.[1] It's a cookbook of sorts. The book contains a recipe for human behaviour – your behaviour. These tech companies know that in order to make money they need to keep us coming back – their business models depend on it.

I know this because I've spent the past decade researching the hidden psychology that some of the most successful companies in the world use to make their products so captivating. For years, I taught future executives at both the Stanford Graduate School of Business and the Hasso Plattner Institute of Design.

In writing *Hooked*, my hope was that startups and socially concerned companies would use this knowledge to design new ways of helping people build better habits. Why should the tech giants keep these secrets to themselves? Shouldn't we use the same psychology that makes video games and social media so engaging to design products to help people live better lives?

Since *Hooked* was published, thousands of companies have used the book to empower their users to build helpful and healthy habits. Fitbod is a fitness app that helps people build better exercise routines. Byte Foods seeks to change people's eating habits with internet-connected pantries that offer locally made fresh meals. Kahoot! builds software to make classroom learning more engaging and fun.*

We want our products to be more user-friendly, easy to navigate, and, yes, habit-forming. Companies making their products more engaging isn't necessarily a problem – it's progress.

But there's also a dark side. As philosopher Paul Virilio wrote, 'When you invent the ship, you also invent the shipwreck.'[2] In the case of user-friendly products and services, what makes some products engaging and easy to use can also make them distracting.

For many people, these distractions can get out of hand, leaving us with a feeling that our decisions are not our own. The fact is, in this day and age if you are not equipped to manage distraction, your brain will be manipulated by time-wasting diversions.

In the next few pages I'll reveal my own struggle with distraction, and how, ironically, I got hooked. But I'll also share how I overcame my struggle and reveal why we are much more powerful than any of the tech giants. As an industry insider, I know their Achilles heel – and soon you will too.

---

*I loved the way Kahoot! and Byte Foods used my book so much, I decided to invest in both companies.

The good news is that we have the unique ability to adapt to such threats. We can take steps right now to retrain and regain our brains. To be blunt, what other choice do we have? We don't have time to wait for regulators to do something and if you hold your breath waiting for corporations to make their products less distracting, well, you're going to pass out.

In the future, there will be two kinds of people in the world: those who let their attention and lives be controlled and coerced by others, and those who proudly call themselves 'indistractable'. By opening this book, you've taken the first step to owning your time and your future.

But you're just getting started. For years you've been conditioned to expect instant gratification. Think of getting to the last page of *Indistractable* as a personal challenge to liberate your mind.

The antidote to impulsiveness is forethought. Planning ahead ensures you will follow through. With the techniques in this book, you'll learn exactly what to do from this day forth to control your attention and choose your life.

1

What's Your Superpower?

I love sweets, I love social media and I love television. However, as much as I love these things, they don't love me back. Overindulging on something sugary after a meal, spending too much time scrolling a feed or bingeing on Netflix until two in the morning were all things I once did with little or no conscious thought – out of habit.

Just as eating too much junk food leads to health problems, the overuse of devices can also have negative consequences. For me, it was the way I prioritised distractions over the most important people in my life. Worst of all was what I let distractions do to my relationship with my daughter. She's our only child and, to my wife and me, the most amazing kid in the world.

One day, the two of us played games from an activity book. We turned to a particular page and answered questions designed to bring dads and daughters closer together. The first activity involved naming each other's favourite things. The next project was to build a paper aeroplane with one of the pages. The third was a question we both had to answer. The question was: 'If you could have any superpower, what would it be?'

I wish I could tell you what my daughter said at that moment, but I can't. I have no idea because I wasn't really there. I was physically in the room, but my mind was elsewhere. 'Daddy?' she said. 'What would your superpower be?'

'Huh?' I grunted. 'Just a second. I just need to respond to this one thing.' I dismissed her as I attended to something on my phone. My eyes were still glued to my screen, tapping away at something that seemed important at the time but could definitely have waited. She went quiet. By the time I looked up, she had gone.

I had just blown a magical moment with my daughter because something on my phone had grabbed my attention. On its own, it was no big deal. But if I told you this was an isolated incident, I'd be lying. This same scene had played out countless times before.

I wasn't the only one putting distractions before people. An early reader of this book told me that when he asked his eight-year-old daughter what her superpower would be, she said she wanted to talk to animals. When asked why, the child said, 'So that I have someone to talk to when you and Mom are too busy working on your computers.'

After finding my daughter and apologising, I decided it was time for a change. At first, I went extreme. Convinced it was all technology's fault, I tried a 'digital detox' and started using an old-school cell phone so I couldn't be tempted to use email, Instagram and Twitter. But I found it too difficult to get around without GPS and meeting addresses saved inside my calendar app. I missed listening to audiobooks while I walked, as well as all the other handy things my smartphone could do.

To avoid wasting time reading too many news articles online, I subscribed to the print edition of a newspaper. A few weeks later, I had a stack of unread papers piled neatly next to me as I watched the news on TV.

In an attempt to stay focused while writing, I bought a 1990s word processor without an internet connection. However, whenever I'd sit down to write, I'd find myself glancing at the bookshelf and soon started flipping through books unrelated to my work. Somehow, I kept getting distracted, even without the tech that I thought was the source of the problem.

Removing online technology didn't work. I'd just replace one distraction with another.

I discovered that living the life we want requires not only doing the *right* things, it also requires that we stop doing the *wrong* things that take us off-track. We all know eating cake is worse for our waistline than having a healthy salad. We agree that aimlessly scrolling our social media feeds is not as enriching as being with real friends in real life. We understand that if we want to be more productive at work, we need to stop wasting time and actually *do* the work. We already know what to do: what we don't know is how to stop getting distracted.

In researching and writing this book over the past five years, and by following the science-backed methods you'll soon learn, I'm now more productive, physically and mentally stronger, better rested, and more fulfilled in my relationships than I've ever been. This book is about what I learned as I developed the most important skill for the twenty-first century. It's about how I became indistractable, and how you can too.

***

The first step is to recognise that distraction starts from within. In Part 1, you'll learn practical ways to identify and manage the psychological discomfort that leads us off-track. However, I steer clear of recommending well-worn techniques like mindfulness and meditation, not because they are not effective for some people, but because these methods have already been written about ad nauseam. If you're reading this book, my guess is that you've already tried those techniques and, like me, found they didn't quite do the trick for you. Instead, we'll take a fresh look at what really motivates our behaviour and learn why time management is pain management. We'll also explore how to make just about any task enjoyable – not in the Mary Poppins way of adding 'a spoonful of sugar', but by cultivating the ability to focus intensely on what we're doing.

Part 2 will look at the importance of making time for the things you really want to do. We'll learn why you can't call something a 'distraction' unless you know what it is distracting you *from*. You'll learn to plan your time with intention, even if you choose to spend it scrolling through celebrity headlines or reading a steamy romance novel. After all, the time you plan to waste is not wasted time.[1]

Part 3 follows with a no-holds-barred examination of the unwanted external triggers that hamper our productivity and diminish our wellbeing. While technology companies use cues like the pings and dings on our phones to hack our behaviour, external triggers are not confined to our digital devices. They're all around us – from cookies beckoning us when we open the kitchen cabinet to a chatty co-worker keeping us from finishing a time-sensitive project.

Part 4 holds the last key to making you indistractable: pacts. While removing external triggers is helpful in keeping distractions *out*, pacts are a proven way of reining ourselves *in*, ensuring we do what we say we're going to do. In this part, we'll apply the ancient practice of precommitment to modern challenges.

Finally, we'll take an in-depth look at how to make your workplace indistractable, raise indistractable kids and foster indistractable relationships. These final chapters will show you how to regain lost productivity at work, have more satisfying relationships with your friends and family, and even be a better lover – all by conquering distraction.

You're welcome to navigate the four steps to becoming indistractable however you like, but I recommend you proceed sequentially through Parts 1 to 4. The four modalities build on each other, with the first step being the most foundational.

If you're the kind of person who likes to learn by example, and you want to see these tactics in action first, feel free to read Parts 5 and on, then come back through the first four parts for a deeper explanation. Also, there's no requirement to adopt each and every technique right away. Some might not fit your current situation and only become useful in the future when you're ready or your circumstances change. But I promise you that by the time you finish this book, you will have discovered several breakthroughs that will change the way you manage distraction forever.

Imagine the incredible power of following through on your intentions. How much more effective would you be at work? How much more time could you spend with your family or doing the things you love? How much happier would you be?

What would life be like if your superpower was to be indistractable?

★ REMEMBER THIS:

- **We need to learn how to avoid distraction.** Living the life we want not only requires doing the right things, but also necessitates *not* doing the things we know we'll regret.
- **The problem is deeper than tech.** Being indistractable isn't about being a Luddite. It's about understanding the real reasons why we do things against our best interests.
- **Here's what it takes.** We can be indistractable by learning and adopting four key strategies.

2

Being Indistractable

The ancient Greeks immortalised the story of a man who was perpetually distracted. We call something that is desirable but just out of reach 'tantalising' after his name. The story goes that Tantalus was banished to the underworld by his father, Zeus, as a punishment.[1] There, he found himself wading in a pool of water, while above his head dangled a branch ripe with fruit ready for the picking.

The curse seems benign, but when Tantalus tried to pluck fruit from the tree, the branch moved away from him, always just out of reach. When he bent down to drink the cool water, it receded so that he could never quench his thirst. Tantalus' punishment was to yearn for things he desired but could never grasp.

You have to give it to the ancient Greeks for their allegories. It's hard to portray a better representation of the human condition. We are constantly reaching for something: more money, more experiences, more knowledge, more status, more *stuff*. The ancient Greeks thought this was just part of the curse of being a fallible mortal and used the story to portray the power of our incessant desires.

Tantalus' curse – forever reaching for something.[2]

TRACTION AND DISTRACTION

Imagine a line that represents the value of everything you do throughout your day. To the right, the actions are positive; to the left, they are negative.

On the right side of the continuum is 'traction', which comes from the Latin *trahere*, meaning to draw or pull. We can think of traction as the actions that draw us towards what we want in life. On the left side is 'distraction', the opposite of traction. With the same Latin root, the word means the 'drawing away of the mind'.[3] Distractions impede us from making progress towards the life we envisage.

All behaviours, both traction and distraction, are prompted by triggers, whether internal or external. Internal triggers cue us from within. When we feel our belly growl, we look for a snack. When we're cold, we find a coat to warm up. And when we're sad, lonely or stressed, we might call a friend or loved one for support.

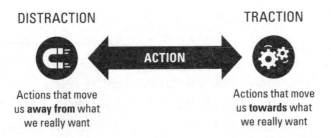

External triggers, on the other hand, are cues in our environment that tell us what to do next, like the pings, dings and rings that prompt us to check our email, answer a phone call or open a news alert. External triggers can also take the form of other people, such as a co-worker who stops by our desk. They can also be objects, like a television set whose mere presence urges us to turn it on.

Whether internal or external triggers prompt us, the resulting action is either aligned with our broader intention (traction), or misaligned (distraction). Traction helps us accomplish goals; distraction leads us away from them.

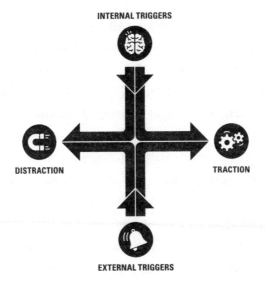

The challenge, of course, is that our world has always been full of things designed to distract us. Today, people find themselves attached to their mobile phones, but these are only the latest potential hindrance. A few decades ago, people complained about the brain-melting power of television.[4] Before that, it was arcade games, the telephone, the pinball machine, comic books and the radio. Even the written word was blamed for creating 'forgetfulness in the learners' souls', according to Socrates.[5] Though some of these things seem dull in comparison to today's enticements, distractions are and always will be a fact of life.

Today's distractions, however, feel different. More data, transferred at faster speeds, enabling ubiquitous access to new content on our devices, means the world can be more distracting. If it's a distraction you seek, it's easier than ever to find.

What is the cost of all that distraction? In 1971, the psychologist Herbert A. Simon wrote presciently, 'the wealth of information means a dearth of something else ... a poverty of attention'.[6]

Researchers tell us attention and focus are the raw materials of human creativity and flourishing.[7] In the age of increased automation, the most sought-after jobs are those that require creative problem-solving, novel solutions and the kind of human ingenuity that comes from focusing deeply on the task at hand.

Socially, we see that close friendships are the bedrock of our psychological and physical health. Loneliness, according to researchers, is more dangerous than obesity.[8] But, of course, we can't cultivate close friendships if we're constantly distracted.

When we consider our children, how can they flourish if they can't concentrate long enough to apply themselves? What example are we setting for them if our loving faces are replaced by the tops of our heads as we constantly stare into our screens?

Let's think back to the tale of Tantalus. What was his curse exactly? Was it never-ending hunger and thirst? Not really. What would have happened to Tantalus if he had just stopped reaching? He was already in hell, after all, and dead people don't *need* food and water last time I checked.

The curse is not that Tantalus spends all eternity reaching for things just out of reach, but rather his obliviousness to the greater folly of his actions. Tantalus' curse was his blindness to the fact he didn't need those things in the first place. That's the real moral of the story.

Tantalus' curse is also our curse. We are compelled to reach for things we supposedly need but really don't. We don't *need* to check our email right this second; we don't *need* to give in to some other distraction, no matter how much we feel we must.

Fortunately, we, unlike Tantalus, can step back from our desires, recognise them for what they are and do something about them. We want companies to innovate and solve our evolving needs, yet we must also ask whether better products bring out the best in us. Though it's not our fault distractions exist (as they always have), managing them is our responsibility.

Being indistractable means striving to do what you say you will do. Indistractable people are as honest with themselves as they are with others.

If you care about your work, your family and your physical and mental wellbeing, you must learn how to become indistractable. The four-part Indistractable Model is a tool for seeing and interacting with the world in a new way. It will serve as your map for controlling your attention and choosing your life.

> ## ⚑ REMEMBER THIS:
>
> - **Distraction stops us from achieving our goals.** It is any action that moves you away from what you really want.
> - **Traction leads you closer to your goals.** It is any action that moves you towards what you really want.
> - **Triggers prompt both traction and distraction.** External triggers prompt you to action with cues *in your environment*. Internal triggers prompt you to action with cues *within you*.

THE INDISTRACTABLE MODEL

Master
INTERNAL TRIGGERS

Prevent
DISTRACTION
With Pacts

Make Time
For **TRACTION**

Hack Back
EXTERNAL TRIGGERS

These four steps are your guide to becoming indistractable.

Part 1

Master Internal Triggers

3

What Motivates Us Really?

Dr Zoë Chance had earned a doctorate from Harvard and taught at the Yale School of Management when she made a shocking revelation to the crowded TEDx audience.[1] 'I'm coming clean today, telling this story for the very first time in its raw, ugly detail,' she said. 'In March of 2012 ... I purchased a device that would slowly begin to ruin my life.'

At Yale, Chance taught future executives the secrets of changing consumer behaviour. Despite the class's title, 'Mastering Influence and Persuasion', Chance's confession revealed that she herself was not immune to manipulation. What began as a research project turned into a mindless compulsion.

Chance stumbled upon a product that typified many of the persuasion techniques she taught in her class. She tells me, 'We kept saying, "Oh, this is brilliant. These guys are geniuses." They've actually used every motivational tool we could possibly think of.'[2]

Naturally, Chance had to try it out for herself and signed up to be the first guinea pig in her research experiment. Little did she know how the product would manipulate her mind and body. 'I really, really, truly could not stop, and it took me a long time to realise it was a problem,' she says now.

It's easy to understand why Chance stayed in denial for so long. The product she became dependent on was not a prescription pill or street drug – it was a pedometer. More specifically, it was the Striiv Smart Pedometer, made by a Silicon Valley start-up founded one year earlier. Chance is quick to mention that the Striiv is no ordinary pedometer. 'They market it as a "personal trainer in your pocket",' she says. 'No! It is Satan in your pocket!'

As a company founded by former video game designers, Striiv utilises behavioural design tactics to compel customers to be more physically active. Users of the pedometer are tasked with challenges as they accrue points for walking. They can compete with other players and view their relative rankings on tournament-style leaderboards. The company also couples the step counter with a smartphone app called MyLand, where players can exchange points to build virtual worlds online.

Clearly, these tricks had cast their spell on Chance. In fact, she found herself incessantly pacing to keep accumulating steps and points. 'I would come home, and while I was eating, or while I was reading, or while I was eating and reading at the same time, or while my husband was trying to talk to me, I would be going in this circuit between the living room and the kitchen and the dining room and the living room and the kitchen and the dining room.'

Unfortunately, all that walking, much of it in circles, started taking its toll. She had less time for her family and friends. 'The only [person] I was getting closer to,' she admits, '[was] my colleague Ernest, who also had a Striiv, so we could set challenges and compete with each other.'

Chance was obsessed. 'I was creating spreadsheets to optimise and track – not my exercising, but my virtual transactions in a virtual world that existed on a Striiv device.' Not only was her obsession sucking time away from her work and other priorities, it also began to cause her physical harm. 'When I was using the Striiv, I was going 24,000 steps a day'.

Chance recalls how at the end of one particularly active day, she received a tempting offer from her Striiv. 'It was midnight, and I was brushing my teeth and getting ready for bed, when this pop-up challenge showed up. It said, "We'll give you triple the points if you just climb twenty stairs!"'

Chance quickly realised she could complete the challenge in about a minute by walking down and up her basement staircase twice. After completing the challenge, she received another message, encouraging her to climb another forty steps for triple points. She thought, 'Yes, of course! It's a good deal!' and quickly walked an additional four flights.

The incessant walking did not stop there. For the next two hours – from midnight until two in the morning – the professor stomped up and down her basement staircase as if possessed by some strange mind-controlling power. When she finally came to a standstill, she realised she had climbed more than 2,000 stairs. That's more than the 1,872 required to climb the Empire State Building.

As she walked up and down her stairs in the middle of the night, she felt unable to stop. Under the influence of the Striiv Smart Pedometer, Zoë Chance had turned into a fitness zombie.

On the surface, her story is a textbook case study of how something as seemingly healthy as a pedometer can mutate into a harmful distraction. Once I'd learned about Chance's strange obsession with her fitness tracker, I wanted to know more. But first, I needed to understand better what really drove her behaviour.

***

For hundreds of years, we've believed that motivation is driven by reward and punishment. As Jeremy Bentham, the English philosopher and founder of modern utilitarianism, put it, 'Nature has placed mankind under the governance of two sovereign masters, pain and pleasure.'[3] The reality, however, is even simpler than that: motivation has less to do with pleasure than was once thought.

Even when we think we're seeking pleasure, we're actually driven by the desire to free ourselves from the pain of wanting.

Epicurus, the ancient Greek philosopher, put it best: 'By pleasure,' he wrote, 'we mean the absence of pain in the body and of trouble in the soul.'[4] Simply put, the drive to relieve discomfort is the root cause of all our behaviour, while everything else is a proximate cause.

Consider the game of pool. What makes the coloured balls go into the pockets? Is it the white cue ball, the cue, or the player's actions? We understand that while the white cue ball and cue are necessary, the root cause is the player using the tools of the game. The white cue ball and cue aren't the root causes; they are the proximate causes of the resulting event.[5]

In the game of life it's often far harder to see the root cause of things. When we're passed over for a promotion, we might blame that cunning co-worker for taking our job instead of reflecting on our lack of qualifications and initiative. When we get into a fight with our spouse over a toilet seat left up, we might blame the conflict on one tiny incident instead of acknowledging years of unresolved issues. And when we scapegoat our political and ideological opponents for the world's troubles, we choose not to seek to understand the deeper systemic reasons behind the problems.

These proximate causes have something in common – they help us deflect responsibility on to something or someone else. It's not that the cue ball and the cue itself don't factor into the equation, just like the co-worker or toilet seat, but they certainly aren't entirely responsible for the outcome. Without understanding and tackling root causes, we're stuck being the helpless victims in a tragedy of our own creation.

The distractions in our lives are the result of the same forces at play – they are proximate causes that we think are to blame, while the root causes stay hidden. We tend to blame things like television, junk food, social media, cigarettes and video games – but these are all proximate causes of our distraction.

Solely blaming a smartphone for causing distraction is just as flawed as blaming a pedometer for making someone climb too many stairs.

Unless we deal with the root causes of our distraction, we'll continue to find ways to distract ourselves. Distraction, it turns out, isn't about the distraction itself; rather, it's about how we respond to it.

***

Over several email exchanges Zoë Chance let me in on the dark truths that drove her extreme behaviours, which she hadn't revealed in her TEDx talk. 'My addiction to Striiv coincided with one of the most stressful periods in my life,' she tells me.[6] 'I was just going on the market to look for a job as a rookie marketing professor: a gruelling, months-long process involving tremendous uncertainty.' She continues, 'It's not uncommon for academics on the job market to experience physical symptoms of stress. I was losing hair, losing sleep and getting heart palpitations. I felt like I was going crazy, and that I had to hide it from everyone.'

Chance was also hiding a secret about her marriage: her husband was a marketing professor, too, meaning the couple 'needed to find a joint appointment, either for [her] at his school, or for both of [them] at another. Marketing departments are small, and joint appointments rare as hen's teeth.'

Further complicating matters, her marriage was falling apart. 'I didn't know whether my husband and I would be together or not, but because the best-case scenario would be that we worked things out, stayed married and I got a job at his university, we didn't want anyone at his university to know we might get divorced, since then they'd be less likely to offer me a job.'

Chance felt stuck. 'I knew that even my best efforts couldn't guarantee a good outcome for either my marriage or the job market, and in hindsight, I can see that Striiv gave me something I could control and succeed at.' During this particularly difficult time in her life, she says she used her Striiv as a coping device. 'It was an escape from reality,' she now admits.

Most people don't want to acknowledge the uncomfortable truth that distraction is always an unhealthy escape from reality. How we deal with uncomfortable internal triggers determines whether we pursue healthful acts of traction or self-defeating distractions.

For Chance, racking up Striiv points provided the escape she was looking for. For other people, the escape comes from checking social media, spending more time in the office, watching television, or, in some cases, drinking or taking hard drugs.

If you're trying to escape the pain of something as serious as impending divorce, the real problem is not your pedometer; without dealing with the discomfort driving the desire for escape, we'll continue to resort to one distraction or another.

Only by understanding our pain can we begin to control it and find better ways to deal with negative urges.

Fortunately, Dr Chance was able to come to this realisation herself. First, she focused on the real source of discomfort in her life, narrowing it down to the internal triggers she was trying to escape. Though she and her husband did eventually separate, she says she's in a much better place in her life now. Professionally, she acquired a full-time post at Yale, where she still teaches today. She has also found better ways to stay healthy and in control of her time, scheduling regular fitness activities instead of letting her pedometer rule over her.

Though overcoming her obsession was a positive step for Chance, the Striiv pedometer won't be the last distraction in her life. But by pinpointing the root cause, rather than blaming the proximate, she'll be better able to address the real issue next time. When used together, the strategies and techniques you're about to learn in this section work both immediately and in the long term to help you deal with the real source of distraction.

⚑ REMEMBER THIS:

- **Understand the root cause of distraction.** Distraction is about more than your devices. Separate proximate causes from the root cause.

- **All motivation is a desire to escape discomfort.** If a behaviour was previously effective at providing relief, we're likely to continue using it as a tool to escape discomfort.

- **Anything that stops discomfort is potentially addictive, but that doesn't make it irresistible.** If you know the drivers of your behaviour, you can take steps to manage them.

4

Time Management Is Pain Management

At first I didn't want to believe the inconvenient truth behind what really drives distraction. But after digesting the scientific literature, I had to face the fact that the motivation for diversion originates within us. As is the case with all human behaviour, distraction is just another way our brains attempt to deal with pain. If we accept this fact, it makes sense that the only way to handle distraction is by learning to handle discomfort.

If distraction costs us time, then time management is pain management.

But where does our discomfort come from? Why are we perpetually restless and unsatisfied? We live in the safest, healthiest, most well-educated, most democratic time in human history,[1] and yet some part of the human psyche causes us to constantly look for an escape from things stirring inside us. As the eighteenth-century English writer Samuel Johnson put it, 'My life is one long escape from myself.'[2] Me too, brother, me too.

Thankfully, we can take solace in knowing we are hardwired for this sort of dissatisfaction. Sorry to say, but the odds are that you and I are never going to be fully satisfied with our lives.

Sporadic bouts of joy, sure. Occasional euphoria? Yes. Singing 'Happy' by Pharrell Williams in your underwear once in a while? OK, who hasn't? But the sustained 'happily ever after' sort of satisfaction you see in the movies? Forget it. It's a myth. That sort of happiness is designed never to last for long. Aeons of evolution gave you a brain in a near-constant state of discontentment.

We're wired this way for a simple reason: as a study published in *Review of General Psychology* notes, 'If satisfaction and pleasure were permanent, there might be little incentive to continue seeking further benefits or advances.'[3] In other words, feeling contented wasn't good for the species. Our ancestors worked harder and strove further because they evolved to be perpetually perturbed, and so we remain today.

Unfortunately, the same evolutionary traits that helped our kin survive by driving them to constantly do more can conspire against us today.

Four psychological factors make satisfaction temporary.

Let's begin with the first factor: boredom. The lengths people will go to avoid boredom is shocking, sometimes literally so. A 2014 study published in *Science* asked participants to sit in a room and think for fifteen minutes.[4] The room was empty except for a device that allowed participants to give themselves a mild but painful electric shock. 'Why would anyone want to do that?' you might ask.

When asked beforehand, every participant in the study said they would pay to avoid getting an electric shock. However, when left alone in the room with the machine and nothing else to do, 67 per cent of men and 25 per cent of women gave themselves electric shocks, and many did so multiple times.

The authors conclude their paper by saying, 'People prefer doing to thinking, even if what they are doing is so unpleasant that they would normally pay to avoid it. The untutored mind does not like to be alone with itself.' It's no surprise, therefore, that most of the top twenty-five websites in America sell escape from our daily drudgery, whether through shopping, celebrity gossip or bite-sized doses of social interaction.[5]

The second psychological factor driving us to distraction is negativity bias, 'a phenomenon in which negative events are more salient and demand attention more powerfully than neutral or positive events'.[6] As the author of one study concluded, 'it appears to be a basic, pervasive fact of psychology that bad is stronger than good'.[7] Such pessimism begins very early in life. Babies begin to show signs of negativity bias at just seven months old, suggesting this tendency is inborn.[8] As further evidence, researchers believe we tend to have an easier time recalling bad memories than good ones. Studies have found people are more likely to recall unhappy moments in their childhood, even if they would describe their upbringing as generally happy.[9]

Negativity bias almost certainly gave us an evolutionary edge. Good things are nice, but bad things can kill you, which is why we pay attention to and remember the bad stuff first. Useful, but what a bummer!

The third factor is rumination, our tendency to keep thinking about bad experiences. If you've ever chewed over something in your mind that you did, or someone did to you, over and over again, seemingly unable to stop thinking about it, you've experienced rumination. This 'passive comparison of one's current situation with some unachieved standard' can manifest itself in self-critical thoughts such as, 'Why can't I handle things better?'[10] As one study notes, 'By reflecting on what went wrong and how to rectify it, people may be able to discover sources of error or alternative strategies, ultimately leading to not repeating mistakes and possibly doing better in the future.'[11] Another potentially useful trait – but, boy, can it make us miserable.

Boredom, negativity bias and rumination can each drive us to distraction. But a fourth factor may be the cruellest of all. Hedonic adaptation – the tendency to return quickly to a baseline level of satisfaction no matter what happens to us in life – is Mother Nature's bait-and-switch. All sorts of life events we think would make us happier actually don't, or at least they don't for long.

For instance, people who have experienced extreme good fortune, such as winning the lottery, have reported that things they had previously enjoyed lost their lustre,[12] effectively returning them to their previous levels of satisfaction. As David Myers writes in *The Pursuit of Happiness*, 'Every desirable experience – passionate love, a spiritual high, the pleasure of a new possession, the exhilaration of success – is transitory.'[13] Of course, as with the other three factors, there are evolutionary benefits to hedonic adaptation. As the author of one study explains, 'new goals continually capture one's attention, one constantly strives to be happy without realizing that in the long run such efforts are futile.'[14]

Cue the sad trombone music. Is futility our fate? Absolutely not. As we've learned, dissatisfaction is an innate power that can be channelled to help us make things better in the same way it served our prehistoric relatives.

Dissatisfaction and discomfort dominate our brain's default state, but we can use them to motivate us instead of defeat us.

Without our species' perpetual disquietude, we would be much worse off – possibly extinct; it is our dissatisfaction that propels us to do everything we do, including to hunt, seek, create and adapt. Even selfless acts, like helping someone, are motivated by our need to escape feelings of guilt and injustice. Our insatiable desire to reach for more is what drives us to overturn despots, it's what pushes the invention of world-changing and life-saving technologies and it's the unseen fuel that launches our ambitions to travel beyond our planet and explore the cosmos.

Dissatisfaction is responsible for our species' advances and its faults. To harness its power, we must disavow the misguided idea that if we're not happy we're not normal – exactly the opposite is true. While this shift in mindset can be jarring, it can also be incredibly liberating.

It's good to know that feeling bad isn't actually bad; it's exactly what survival of the fittest intended.

From that place of acceptance, we stand a chance of avoiding the pitfalls of our psyches. We can recognise pain and rise above it, which is the first step on the road to becoming indistractable.

> ★ **REMEMBER THIS:**
>
> - **Time management is pain management.** Distractions cost us time, and, like all actions, they are spurred by the desire to escape discomfort.
> - **Evolution favours dissatisfaction over contentment.** Our tendencies towards boredom, negativity bias, rumination and hedonic adaptation conspire to make sure we're never satisfied for long.
> - **Dissatisfaction is responsible for our species' advancements as much as its faults.** It is an innate power that can be channelled to help us make things better.
> - **If we want to master distraction, we must learn to deal with discomfort.**

5

Deal with Distraction from Within

Dr Jonathan Bricker has spent his career helping people manage the kind of discomfort that leads not only to distraction, but also to disease. As a psychologist at the Fred Hutchinson Cancer Research Center in Seattle, his work has been proven to effectively reduce the risk of cancer by changing patient behaviour. Bricker writes, 'Most people don't think of cancer as a behavioural problem, but whether it's quitting smoking or losing weight or exercising more, there are some definitive things you can do to reduce your risk and thereby live a longer and higher-quality life.'[1]

Bricker's approach involves harnessing the power of imagination to help his patients see things differently. His work shows how learning certain techniques as part of acceptance and commitment therapy (ACT) can disarm the discomfort that so often leads to harmful distractions.

Bricker decided to focus his efforts on stopping smoking and developed an app to deliver ACT over the internet. Though he uses ACT specifically to help people quit smoking, the principles of the programme have been shown to effectively reduce many types of urges. At the heart of the therapy is learning to notice and accept one's cravings and to handle them in a healthy way. Instead of suppressing urges, ACT prescribes a method for stepping back, noticing, observing and finally letting the desire disappear naturally.

But why not simply fight our urges? Why not 'just say no'?

It turns out mental abstinence can backfire.

Fyodor Dostoevsky wrote in 1863, 'Try to pose for yourself this task: not to think of a polar bear, and you will see that the cursed thing will come to mind every minute.'[2] One hundred and twenty-four years later, the social psychologist Daniel Wegner put Dostoevsky's claim to the test.

In a study, participants who were told to avoid thinking of a white bear for five minutes did so on average once per minute, just as Dostoevsky predicted. But there was more to Wegner's study. When the same group was told to try to conjure the white bear, they did so much more often than a group who hadn't been asked to suppress the thought. 'The results suggested that suppressing the thought for the first five minutes caused it to "rebound" even more prominently into the participants' minds later,' according to an article in *Monitor on Psychology*.[3] Wegner later dubbed this tendency 'ironic process theory' to explain why it's so difficult to tame intruding thoughts.

The 'ironic' part of the ironic process theory is the fact that relief of the tension of wanting makes relieving it all the more rewarding, and therefore habit-forming.

An endless cycle of resisting, ruminating, and finally giving in to the desire perpetuates the cycle and quite possibly drives many of our unwanted behaviours.

For example, many smokers believe the chemical nicotine causes their cravings. They're certainly not wrong, but they're not completely right either. Nicotine produces distinct physical sensations. However, a fascinating study involving flight attendants demonstrated how smoking cravings might have much less to do with nicotine than we once thought.

Two groups of flight attendants who smoked were sent on two separate flights from Israel. One group was sent on a three-hour flight to Europe, while the other group travelled to New York, a ten-hour flight. All the smokers were asked by the researchers to rate their level of cravings at set time intervals before, during and after the flight.[4] If cravings were driven solely by the effect of nicotine on the brain, one would expect that both groups would report strong urges after the same number of minutes had elapsed since their last cigarette; the more time passed, the more their brains would chemically crave nicotine. But that's not what happened.

When the flight attendants flying to New York were above the Atlantic Ocean, they reported weak cravings. Meanwhile, at the exact same moment, the cravings of their colleagues who had just landed in Europe were at their strongest. What was going on?

The New York-bound flight attendants knew they could not smoke in mid-flight without being fired. Only later, when they approached their destination, did they report the greatest desire to smoke. It appeared that the duration of the trip and the time since their last cigarette didn't affect the level of the flight attendants' cravings.

What affected their desire was not how much time had passed after a smoke, but how much time was left *before* they could smoke again.[5] If, as this study suggests, a craving for something as addictive as nicotine can be manipulated in this way, why can't we trick our brains into mastering other unhealthy desires? Thankfully, we can!

You'll notice that throughout this book I cite smoking cessation and drug addiction research. I do this for two reasons: first, though studies show very few people are pathologically 'addicted' to distractions like the internet,[6] tech overuse can look to many like an addiction. Second, I wanted to make the point that if these well-established techniques are effective at stopping physical dependencies on nicotine and other substances, then they can *certainly* help us control cravings for distraction. After all, we're not injecting Instagram or freebasing Facebook.

Certain desires can be modulated, if not completely mitigated, by how we think about our urges. In the following chapters, we'll learn how to think differently about three things: the internal trigger, the task and our temperament.

★ REMEMBER THIS:

- **Without techniques for disarming temptation, mental abstinence can backfire.** Resisting an urge can trigger rumination and make the desire grow stronger.
- **We can manage distractions that originate from within by changing how we think about them.** We can reimagine the trigger, the task and our temperament.

6

Reimagine the Internal Trigger

While we can't control the feelings and thoughts that pop into our heads, we can control what we do with them. Bricker's work using acceptance and commitment therapy in smoking cessation programmes suggests we shouldn't keep telling ourselves to stop thinking about an urge; instead, we must learn better ways to cope. The same applies to other distractions like checking our phones too much, eating junk food or excessive shopping. Rather than trying to fight the urge, we need new methods to handle intrusive thoughts. The following four steps help us do just that:

Step 1: Look for the discomfort that precedes the distraction, focusing in on the internal trigger

A common problem I have while writing is the urge to google something. It's easy to justify this bad habit as 'doing research' but deep down I know it's often just a diversion from difficult work. Bricker advises focusing on the internal trigger that precedes the unwanted behaviour, like 'feeling anxious, having a craving, feeling restless, or thinking you are incompetent'.[1]

Step 2: Write down the trigger

Bricker advises writing down the trigger, whether or not you subsequently give in to the distraction, using 'a journal, a piece of paper, a chart, or an app'. He recommends noting the time of day, what you were doing and how you felt when you noticed the internal trigger that led to the distracting behaviour 'as soon as you are aware of the behaviour', because it's easier at that point to remember how you felt. I've included a Distraction Tracker at the back of this book on which you can note the triggers you experience throughout the day. You can download and print additional copies at NirAndFar.com/Indistractable; keep it handy for easy access.

According to Bricker, while people can easily identify the external trigger, 'it takes some time and trials to begin noticing those all-important inside triggers'. He recommends discussing the urge as if you were an observer, telling yourself something like, 'I'm feeling that tension in my chest right now. And there I go, trying to reach for my iPhone.' The better we are at noticing the behaviour, the better we'll be at managing it over time. 'The anxiety goes away, the thought gets weaker or [is] replaced by another thought.'

Step 3: Explore your sensations

Bricker then recommends getting curious about that sensation. For example, do your fingers twitch when you're about to be distracted? Do you get a flurry of butterflies in your stomach when you think about work when you're with your family? What does it feel like when the feelings crest and then subside? Bricker encourages staying with the feeling before acting on the impulse.

When similar techniques were applied in a smoking cessation study, the participants who had learned to acknowledge and explore their cravings managed to quit at twice the rate of those in the American Lung Association's best-performing cessation programme.[2]

One of Bricker's favourite techniques is the 'leaves on a stream' method. When feeling the uncomfortable internal trigger to do something you'd rather not, 'imagine you are seated beside a gently flowing stream', he says. 'Then imagine there are leaves floating down that stream. Place each thought in your mind on each leaf. It could be a memory, a word, a worry, an image. And let each of those leaves float down that stream, swirling away, as you sit and just watch.'

Step 4: Beware of liminal moments

Liminal moments are transitions from one thing to the other throughout our day. Have you ever picked up your phone while waiting for a traffic light to change, then found yourself still looking at your phone while driving? Or opened a tab in your web browser, got annoyed by how long it's taking to load and opened up another page while you waited? Or looked at a social media app while walking from one meeting to the next, only to keep scrolling when you got back to your desk? There's nothing wrong with any of these actions per se. Rather, what's dangerous is the fact that by doing them 'for just a second' we're likely to do things we later regret, like getting off-track for half an hour or getting into a car accident.

A technique I've found particularly helpful for dealing with this distraction trap is the 'ten-minute rule'.[3] If I find myself wanting to check my phone as a pacification device when

I can't think of anything better to do, I tell myself it's fine to give in, but not right now. I have to wait just ten minutes. This technique is effective at helping me deal with all sorts of potential distractions, like Googling something rather than writing, eating something unhealthy when I'm bored, or watching another episode on Netflix when I'm 'too tired to go to bed'.

This rule allows time to do what some behavioural psychologists call 'surfing the urge'.[4] When an urge takes hold, noticing the sensations and riding them like a wave – neither pushing them away nor acting on them – helps us cope until the feelings subside.

Surfing the urge, along with other techniques to bring attention to the craving, has been shown to reduce the number of cigarettes smokers consumed when compared to those in a control group who didn't use the technique.[5] If we still want to perform the action after ten minutes of urge surfing, we're free to do it, but that's rarely still the case. The liminal moment has passed, and we're able to do the thing we really wanted to do.

Techniques like surfing the urge and thinking of our cravings as leaves on a stream are mental skill-building exercises that can help us stop impulsively giving in to distractions. They recondition our minds to seek relief from internal triggers in a reflective rather than a reactive way. As Oliver Burkeman wrote in the *Guardian*, 'It's a curious truth that when you gently pay attention to negative emotions, they tend to dissipate – but positive ones expand.'[6]

We've considered how we might reimagine the internal triggers; next we'll learn how to reimagine the task we're trying to stay focused on.

REMEMBER THIS:

- **By reimagining an uncomfortable internal trigger, we can disarm it.**
- **Step 1.** Look for the emotion preceding distraction.
- **Step 2.** Write down the internal trigger.
- **Step 3.** Explore the negative sensation with curiosity instead of contempt.
- **Step 4.** Be extra cautious during liminal moments.

7

Reimagine the Task

Ian Bogost studies fun for a living. A professor of interactive computing at the Georgia Institute of Technology, Dr Bogost has written ten books, including quirky titles such as *How to Talk About Videogames*, *The Geek's Chihuahua* and, most recently, *Play Anything*. In this latest book, Bogost makes several bold claims that challenge the way we think about fun and play. 'Fun,' he writes, 'turns out to be fun even if it doesn't involve much (or any) enjoyment.'[1] Huh?

Doesn't fun have to feel good? Not necessarily, Bogost says. By relinquishing our notions about what fun should feel like, we open ourselves up to seeing tasks in a new way. He advises that play can be part of any difficult task and, though play doesn't necessarily have to be pleasurable, it can free us from discomfort – which, let's not forget, is the central ingredient driving distraction.

Given what we know about our propensity for distraction when we're uncomfortable, reimagining difficult work as fun could prove incredibly empowering. Imagine how powerful you'd feel if you were able to transform the hard, focused work you have to do into something that felt like play. Is that even possible? Bogost thinks it is, but probably not in the way you think.

Fun and play don't have to make us feel good per se; rather, they can be used as tools to keep us focused.

We've all heard Mary Poppins' advice to add 'a spoonful of sugar' to make a task into a game, yes? Well, Bogost believes Poppins was wrong, saying, 'it recommends covering over drudgery'. As he writes, 'We fail to have fun because we don't take things seriously *enough*, not because we take them so seriously that we'd have to cut their bitter taste with sugar. Fun is not a feeling so much as an exhaust produced when an operator can treat something with dignity.'

Bogost tells us that 'fun is the aftermath of deliberately manipulating a familiar situation in a new way'. The answer, therefore, is to focus on the task itself. Instead of running away from our pain or using rewards like prizes and treats to help motivate us, the idea is to pay such close attention that you find new challenges you didn't see before. Those new challenges provide the novelty to engage our attention and maintain focus when tempted by distraction.

Countless commercially produced distractions, like television or social media, use slot machine-like variable rewards to keep us engaged with a constant stream of newness. But Bogost points out that we can use the same techniques to make any task more pleasurable and compelling.

We can use the same neural hardwiring that keeps us hooked to media to keep us engaged in an otherwise unpleasant task.

Bogost gives the example of mowing his lawn. 'It may seem ridiculous to call an activity like this "fun",' he writes, yet he learned to love it. Here's how: 'First, pay close, foolish, even absurd attention to things.' For Bogost, this meant soaking up as much information as he could about the way grass grows and how to treat it. Then, he created an 'imaginary playground in which the limitations . . . produce[d] meaningful experiences'. He learned about the constraints he had to operate under, including his local weather conditions and what different kinds of equipment can and can't do. Operating under constraints, Bogost says, is the key to creativity and fun. Finding the optimal path for the mower or beating a record time are other ways to create an imaginary playground.

While learning how to have fun cutting grass may seem like a stretch, people find fun in a wide range of activities that you might not find particularly interesting. Consider my local coffee-obsessed barista, who spends a ridiculous amount of time refining the perfect brew, the car buff who toils for countless hours fine-tuning his ride or the crafter who produces sweaters and quilts in the name of 'fun'. If people can have fun doing these activities by choice, what's so crazy about bringing the same kind of mindset to other tasks?

For me, I learned to stay focused on the tedious work of writing books by finding the mystery in my work. I write to answer interesting questions and discover novel solutions to old problems. To quote a classic writer's motto, 'The cure for boredom is curiosity. There is no cure for curiosity.'[2] Today, I write for the fun of it. Of course, it's also my profession, but by finding the fun I'm able to do my work without getting distracted nearly as often.

Fun is looking for the variability in something other people don't notice. It's breaking through the boredom and monotony to discover its hidden beauty.

The great thinkers and tinkerers of history made their discoveries because they were obsessed with the intoxicating draw of discovery – the mystery that pulls us in because we want to know more.

But remember: finding novelty is only possible when we give ourselves the time to focus intently on a task and look hard for the variability. Whether it's uncertainty about our ability to do a task better or faster than last time or coming back to challenge the unknown day after day, the quest to solve these mysteries is what turns the discomfort we seek to escape with distraction into an activity we embrace.

The last step in managing the internal triggers that can lead to distraction is to reimagine our capabilities. We'll start by shattering a common self-defeating belief many of us tell ourselves daily.

★ REMEMBER THIS:

- **We can master internal triggers by reimagining an otherwise dreary task.** Fun and play can be used as tools to keep us focused.
- **Play doesn't have to be pleasurable.** It just has to hold our attention.
- **Deliberateness and novelty can be added to any task to make it fun.**

8

Reimagine Your Temperament

To manage the discomfort that tugs us towards distraction, we need to think of ourselves differently. The way we perceive our temperament, which is defined as 'a person's or animal's nature, especially as it permanently affects their behaviour',[1] has a profound impact on how we behave.

One of the most pervasive bits of folk psychology is the belief that self-control is limited; that by the nature of our temperament we only have so much willpower available to us. Furthermore, the thinking goes, we are liable to run out of willpower when we exert ourselves. Psychologists have a name for this phenomenon: 'ego depletion'.

Not so long ago, my after-work routine looked like this: I'd sit on the couch and veg for hours, keeping company with Netflix and a pint of ice cream (Ben & Jerry's Chocolate Fudge Brownie, to be precise). I knew the ice cream and the sitting weren't good for me, but I justified my actions by telling myself I was 'spent', acting as if my ego were depleted (even if I'd never heard the term). This theory would seem to explain perfectly my after-work indulgences. But is ego depletion real?

In 2011, the psychologist Roy Baumeister of Case Western Reserve University wrote *Willpower: Rediscovering the Greatest Human Strength* with *New York Times* journalist John Tierney.[2] The book quickly became a bestseller. Its authors cited several of Baumeister's studies demonstrating the ego depletion theory, including one notable experiment that showed a seemingly miraculous way to restore willpower: by consuming sugar.[3] The study claimed that participants who had sipped sugar-sweetened lemonade demonstrated increased self-control and stamina on difficult tasks.

Recently, however, scientists have examined the theory more critically and several have soured on the idea. Evan Carter at the University of Miami was one of the first to challenge Baumeister's findings. In a 2010 meta-analysis (a study of studies), Carter looked at nearly 200 experiments that concluded ego depletion was real. Upon closer inspection, however, he identified a 'publication bias', in which studies that produced contradictory evidence were not included.[4] When factoring in their results, he concluded there was no firm evidence supporting the ego depletion theory.[5] Furthermore, some of the more magical aspects of the theory, such as the idea that sugar can increase willpower, have been thoroughly debunked.[6]

What might explain the ego depletion phenomenon? The results of early studies may have been authentic, but it appears the researchers jumped to the wrong conclusions. New studies show that drinking lemonade *can* improve performance, but not for the reason Baumeister believed. Rather, the bump in performance had nothing to do with the sugar in the drink and everything to do with the thoughts in our heads.

In a study conducted by the Stanford psychologist Carol Dweck and her colleagues, published in the *Proceedings of the National Academy of Sciences*, Dweck concluded that signs of ego depletion were observed only in those test subjects who believed willpower was a limited resource.[7] It wasn't the sugar in the lemonade but the belief in its impact that gave participants an extra boost.

People who did not see willpower as a finite resource did not show signs of ego depletion.

Many people still promote the idea of ego depletion, perhaps because they are unaware of the evidence that exists to the contrary. But if Dweck's conclusions are correct, then perpetuating the idea is doing real harm. If ego depletion is essentially caused by self-defeating thoughts and not by any biological limitation, then the idea makes us less likely to accomplish our goals by providing a rationale to quit when we could otherwise persist.

Michael Inzlicht, a professor of psychology at the University of Toronto and the principal investigator at the Toronto Laboratory for Social Neuroscience, offers an alternative view: he believes that willpower is not a finite resource but instead acts like an emotion.[8] Just as we don't 'run out' of joy or anger, willpower ebbs and flows based on what's happening to us and how we feel.

Seeing the link between temperament and willpower through a different lens has profound implications for the way we focus our attention. For one, if mental energy is more like an emotion than fuel in a tank, it can be managed and utilised as such.

A toddler might throw a temper tantrum when refused a toy but will, as he or she ages, gain self-control and learn to ride out bad feelings. Similarly, when we need to perform a difficult task, it's more productive and healthier to believe a lack of motivation is temporary than it is to tell ourselves we're spent and need a break (and maybe some ice cream).

While we can stop believing our willpower is limited, our view of it is just one facet of temperament. Several recent studies have found a strong connection between the way we think about other aspects of human nature and our ability to follow through.

For example, to determine how in control people feel regarding their cravings for cigarettes, drugs or alcohol, researchers administer a standard survey called the Craving Beliefs Questionnaire.[9] The assessment is modified for the participant's drug of choice and presents statements like, 'Once the craving starts ... I have no control over my behaviour', and the cravings 'are stronger than my willpower'.

How people rate these statements tells researchers a great deal, not only about their current state, but also how likely they are to remain addicted. Participants who indicate they feel more powerful as time passes increase their odds of quitting.[10] In contrast, studies of methamphetamine users and cigarette smokers found that those who believed they were powerless to resist were most likely to fall off the wagon after quitting.[11]

The logic isn't surprising, but the extent of the effect is remarkable. A study published in the *Journal of Studies on Alcohol and Drugs* found that individuals who believed they were powerless to fight their cravings were much more likely to drink again.[12]

Addicts' belief regarding their powerlessness was just as significant in determining whether they would relapse after treatment as their level of physical dependence.

Just let that sink in – mindset mattered as much as than physical dependence! What we say to ourselves is vitally important. Labelling yourself as having poor self-control actually leads to less self-control.[13] Rather than telling ourselves we failed because we're somehow deficient, we should offer self-compassion by speaking to ourselves with kindness when we experience setbacks.

Several studies have found people who are more self-compassionate experience a greater sense of wellbeing. A 2015 review of seventy-nine studies looking at the responses of over 16,000 volunteers found that people who have 'a positive and caring attitude … toward her- or himself in the face of failures and individual shortcomings' tend to be happier.[14] Another study found that people's tendency to self-blame, along with how much they ruminated on a problem, could almost completely mitigate the most common factors associated with depression and anxiety.[15] An individual's level of self-compassion had a greater effect on whether they would develop anxiety and depression than all the usual things that tend to screw up people's lives, like traumatic life events, a family history of mental illness, low social status, loneliness or a lack of social support.

The good news is that we can change the way we talk to ourselves in order to harness the power of self-compassion. This doesn't mean we won't mess up; we all do. Everyone struggles with distraction from one thing or another. The important thing is to take responsibility for our actions without heaping on the toxic guilt that makes us feel even worse and can, ironically, lead us to seek even more distraction, in order to escape the pain of shame.

Self-compassion makes people more resilient to letdowns by breaking the vicious cycle of stress that often accompanies failure.

If you find yourself listening to the little voice in your head that sometimes bullies you around, it's important to know how to respond. Instead of accepting what the voice says or arguing with it, remind yourself that obstacles are part of the process of growth. We don't get better without practice, which can be clumsy and difficult at times.

A good rule of thumb is to talk to yourself the way you might talk to a friend. Since we know so much about ourselves, we tend to be our own worst critics, but if we talk to ourselves the way we'd help a friend, we can see the situation for what it really is. Telling yourself, 'This is what it's like to get better at something' and 'You're on your way' is a healthier way to handle self-doubt.

***

Reimagining the internal trigger, the task and our temperament are powerful and proven ways to deal with distractions that start within us. We can cope with uncomfortable internal triggers by reflecting on, rather than reacting to, our discomfort. We can reimagine the task we're trying to accomplish by looking for the fun in it and focusing on it more intensely. Finally, and most importantly, we can change the way we see ourselves to get rid of self-limiting beliefs.

If we believe we're short on willpower and self-control, then we will be. If we decide we're powerless to resist temptation, it becomes true. If we tell ourselves we're deficient by nature, we'll believe every word.

Thankfully, you don't have to believe everything you think; you are only powerless if you think you are.

★ REMEMBER THIS:

- **Reimagining our temperament can help us manage our internal triggers.**
- **We don't run out of willpower.** Believing we do makes us less likely to accomplish our goals, by providing a rationale to quit when we could otherwise persist.
- **What we say to ourselves matters.** Labelling yourself as having poor self-control is self-defeating.
- **Practise self-compassion.** Talk to yourself the way you'd talk to a friend. People who are more self-compassionate are more resilient.

Part 2

Make Time for Traction

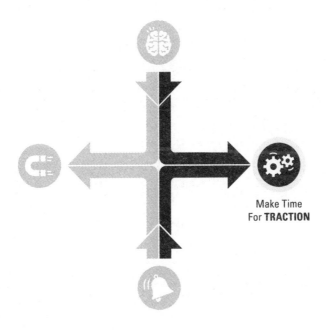

Make Time
For **TRACTION**

9

Turn Your Values into Time

Traction draws you towards what you want in life, while distraction pulls you away. In Part 1, we learned ways to cope with the internal triggers that can drive us to distraction and how to reduce the sources of discomfort: if we don't control our impulse to escape uncomfortable feelings, we'll always look for quick fixes to soothe our pain.

The next step is to find ways to make traction more likely, starting with how we spend our time. The German philosopher Johann Wolfgang von Goethe believed the way someone spent their time could tell you everything. 'If I know how you spend your time,' he wrote, 'then I know what might become of you.'[1]

Think of all the ways people steal your time. Seneca, the Roman Stoic philosopher, wrote, 'People are frugal in guarding their personal property; but as soon as it comes to squandering time, they are most wasteful of the one thing in which it is right to be stingy.'[2] Though Seneca was writing more than 2,000 years ago, his words are just as applicable today. As he noted, people protect their property in all sorts of ways – locks, security systems and storage units – but most do little to protect their time.

A study by PPAI Research found only a third of Americans keep a daily schedule,[3] which means the vast majority wake up every morning with no formal plans. Our most precious asset – our time – is unguarded, just waiting for someone to steal it. If we don't plan our day, someone else will.[4]

When it comes to planning our schedule, where do we begin? The standard approach is to make a to-do list. We write down all the things we want to do and hope we'll find the time throughout the day to do them. Unfortunately, this method has some serious flaws. Anyone who has tried keeping such a list knows many tasks tend to get pushed from one day to the next, and the next. Instead of starting with *what* we're going to do, we should begin with *why* we're going to do it. And to do that, we must begin with our values.

According to Russ Harris, author of *The Happiness Trap*, values are 'how we want to be, what we want to stand for, and how we want to relate to the world around us'.[5] They are attributes of the person we want to be. For example, they may include being an honest person, being a loving parent or being a valued part of a team. We never *achieve* our values any more than finishing a painting would let us *achieve* being creative. Values are not end goals; they are guidelines for our actions.

Though some values carry over into all facets of life, most are specific to one area. For example, being a contributing member of a team is something people generally do at work. Being a loving spouse or parent occurs within the context of a family. Being the kind of person who seeks wisdom or physical fitness is something we do for ourselves.

The trouble is, we don't make time for our values. We unintentionally spend too much time in one area of our life at the expense of others. We get busy at work at the expense of living out our values with our family or friends. If we run ourselves ragged caring for our kids, we neglect our bodies, minds and friendships, keeping us from being the person we desire to be. If we chronically neglect our values, we become someone we're not proud of – our life feels out of balance and diminished. Ironically, this ugly feeling makes us more likely to seek distractions to escape our dissatisfaction without actually solving the problem.

Though each of us may subscribe to different values, it's helpful to categorise them into various life domains, a concept that is thousands of years old. The Stoic philosopher Hierocles demonstrated the interconnected nature of our lives by the use of concentric circles.[6] He placed the human mind and body at the centre, followed by family in the next ring, fellow citizens and countrymen next, and all humanity in the outermost ring.

Inspired by his example, I created a way to simplify and visualise the three life domains where we spend our time:

LIFE DOMAINS

WORK

RELATIONSHIPS

YOU

The three life domains: you, relationships and work.

These three domains describe with whom we spend our time. They give us a way to think about how we plan our days so that we can become an authentic reflection of the person we want to be.

Only by setting aside specific times in our schedules for traction (the actions that draw us towards what we want in life) can we turn our backs on distraction. In order to live our values in each of these domains, we must reserve time in our schedules to do so. Without planning ahead, it's impossible even to tell the difference between traction and distraction.

You can't call something a distraction unless you know what it's distracting you from.

I know many of us bristle at the idea of keeping a schedule because we don't want to feel hampered, but, oddly enough, we actually perform better under constraints.[7] This is because limitations give us a structure, while a blank schedule and a mile-long to-do list torment us with too many choices.

The most effective way to make time for traction is through timeboxing. Timeboxing uses a well-researched technique psychologists call 'setting an implementation intention', which is a fancy way of saying 'deciding *what* you're going to do, and *when* you're going to do it'.[8] It's a technique that can be used to make time for traction in each of your life domains.

The goal is to eliminate all white space in your calendar, so you're left with a template for how you intend to spend your time each day.

It doesn't so much matter *what* you do with your time; rather, success is measured by whether you did what you planned to do. It's fine to watch a video, scroll social media, daydream or take a nap, as long as that's what you planned to do. Alternatively, checking work email, a seemingly productive task, is a distraction if it's done when you intended to spend time with your family or work on a presentation. Keeping a timeboxed schedule is the only way to know if you're distracted; if you're not spending your time doing what you'd planned, you're off-track.

To create a weekly timeboxed schedule, you'll need to decide how much time you want to spend on each domain of your life. How much time do you want to spend on yourself, with important relationships, and on your work? Note that 'work' doesn't exclusively mean paid labour. The work domain can include community service, civic engagement in government, and side projects.

How much time in each domain would allow you to be consistent with your values? Start by creating a weekly calendar template for your perfect week. You'll find a blank Schedule Template at the end of the book and a free online tool at NirAndFar.com/Indistractable.

Next, set aside fifteen minutes on your schedule every week to reflect and refine your calendar by asking yourself two questions:

Question 1 (Reflect): 'When in my schedule did I do what I said I would do and when did I get distracted?' Answering this question requires you to look back at the past week. I recommend using the Distraction Tracker found at the back of this book to note down when and why you become distracted, as per Dr Bricker's suggestions of noting your internal trigger from Chapter 6.

If you became distracted as a result of an internal trigger, what strategies will you use to cope the next time it arises? Did an external trigger, like a phone call or a talkative colleague, prompt you to stop doing what you wanted to do? (We'll address tactics to control external triggers in Part 3.) Or was a planning problem the reason you gave in to distraction? In which case, you can look back through your Distraction Tracker to help answer the next question.

Question 2 (Refine): 'Are there changes I can make to my calendar that will give me the time I need to better express my values?' Maybe something unexpected came up, or perhaps there was a problem with how you planned your day. Timeboxing enables us to think of each week as a mini-experiment. The goal is to figure out where your schedule didn't work out in the prior week so you can make it easier to follow the next time around. The idea is to commit to a practice that improves your schedule over time by helping you know the difference between traction and distraction for every moment of the day.

When our lives change, our schedules can, too. But once our schedule is set, the idea is to stick with it until we decide to improve it on the next go-round. Approaching the exercise of making a schedule as a curious scientist, rather than a drill sergeant, gives us the freedom to get better with each iteration.

In this section we'll look at how to make time for traction in the three domains of your life. We'll also discuss how to sync expectations of how you spend your time with the stakeholders in your life, such as family members, co-workers and managers.

Before moving on, consider what your schedule currently looks like. I'm not asking about the things you *did*, but, rather, the things you committed to doing in writing. Is your schedule filled with carefully timeboxed plans, or is it mostly empty? Does it reflect who you are? Are you letting others steal your time or do you guard it as the limited and precious resource it is?

By turning our values into time, we make sure we have time for traction. If we don't plan ahead, we shouldn't point fingers or be surprised that everything becomes a distraction. Being indistractable is largely about making sure you make time for traction each day and eliminating the distraction that keeps you from living the life you want – one that involves taking care of yourself, your relationships and your work.

★ REMEMBER THIS:

- **You can't call something a distraction unless you know what it is distracting you from.** Planning ahead is the only way to know the difference between traction and distraction.
- **Does your calendar reflect your values?** To be the person you want to be, you have to make time to live your values.
- **Timebox your day.** The three life domains of you, relationships and work provide a framework for planning how to spend your time.
- **Reflect and refine.** Revise your schedule regularly, but you must commit to it once it's set.

10

Control the Inputs, Not the Outcomes

LIFE DOMAINS

WORK

RELATIONSHIPS

YOU

In this visual representation of your life, you are at the centre of the three domains. As with everything valuable, you require maintenance and care, which takes time. Just as you wouldn't blow off a meeting with your boss, so you should never bail on appointments you make with yourself. After all, who's more critical to helping you live the kind of life you want than you yourself?

Exercise, sleep, eating healthily and time spent reading or listening to an audiobook are all ways to invest in ourselves. Some people value mindfulness, spiritual connection or reflection, and may want time to pray or meditate. Others value skilfulness and want time alone to practise a hobby.

Taking care of yourself is at the core of the three domains because the other two depend on your health and wellness. If you're not taking care of yourself, your relationships suffer. Likewise, your work isn't its best when you haven't given yourself the time you need to stay physically and psychologically healthy.

We can start by prioritising and timeboxing 'You' time. At a basic level, we need time in our schedules for sleep, hygiene and proper nourishment. While it may sound simple to fulfil these needs, I must admit that before I learned to timebox my day I was guilty of spending many late nights at work, after which I'd quickly grab a double cheeseburger, curly fries and a decadent chocolate shake for dinner – a far cry from the healthy lifestyle I envisaged.

By setting aside time to live out your values in the 'You' domain, you will not only be able to visualise the qualities of the person you want to be by reflecting on your calendar, but you will also be much more likely to do what you promised yourself you would do.

You might be thinking, 'It's all well and good to schedule time for ourselves, but what happens when we don't accomplish what we want to, despite making the time?' Let's say, for example, you aren't getting enough sleep. You've probably read many articles about the importance of rest, so I'm not going to rehash the point other than to say the research is unequivocal – we need quality sleep.[1]

A few years ago I started waking up at three o'clock every morning. I'd toss and turn, disappointed that I wasn't following through on my plan to get seven to eight hours of shut-eye. It was on my schedule, so why wasn't I asleep? It turns out that sleeping wasn't completely under my control. I couldn't help the fact that my body chose to wake me up, but I could control what I did in response.

At first, I did what many of us do when things don't go as planned – I freaked out. I'd lie in bed, thinking about how bad it was that I wasn't sleeping and how I was likely to feel groggy the next day, and then I'd start thinking of all the things I had to do the next day. I'd mull over these thoughts until I could think of nothing else. Ironically, I wasn't falling asleep again because I was worried about not falling asleep again – a common cause of insomnia.[2]

Once I realised my rumination was itself a distraction, I began to deal with it in a healthier manner. Specifically, if I woke up I'd repeat a simple mantra: 'The body gets what the body needs.' That subtle mindset shift reduced the pressure by no longer making sleep a requirement. My job was to provide my body with the proper time and place to rest; what happened next was out of my control.

I started to think of waking up in the middle of the night as a chance to read on my Kindle and stopped worrying about when I'd fall asleep again.* I assured myself that if I wasn't tired enough to fall asleep right at that moment, it was because my body had already got enough rest. I let my mind relax without worrying.

*The Kindle e-reader is less harmful to sleep than other devices. Anne-Marie Chang, Daniel Aeschbach, Jeanne F. Duffy and Charles A. Czeisler, 'Evening Use of Light-Emitting EReaders Negatively Affects Sleep, Circadian Timing, and Next-Morning Alertness', *Proceedings of the National Academy of Sciences* 112, no. 4 (27 January 2015): 1232, https://doi.org/10.1073/pnas.1418490112.

You see where this is leading, don't you? Once my rumination stopped, so did my sleepless nights. I soon started regularly falling back asleep within minutes.

There's an important lesson here that goes well beyond how to get enough sleep. The takeaway is that, when it comes to our time, we should stop worrying about outcomes we can't control and instead focus on the inputs we can. The positive results of the time we spend doing something is a hope, not a certainty.

The one thing we control is the time we put into a task.

Whether or not I'm able to fall asleep at any given moment or whether a breakthrough idea for my next book comes to me when I sit down at my desk isn't entirely up to me, but one thing is certain: I won't do what I want to do if I'm not in the right place at the right time, whether that's in bed when I want to sleep or at my desk when I want to do good work. Not showing up guarantees failure.

We tend to think we can solve our distraction problems by trying to get more done each minute, but the real problem is more often not giving ourselves time to do what we say we will. By timeboxing 'You' time and faithfully following through, we keep the promises we make to ourselves.

★ REMEMBER THIS:

- **Schedule time for yourself first.** You are at the centre of the three life domains. By not allocating time for yourself, the other two domains suffer.
- **Show up when you say you will.** You can't always control what you get out of time you spend, but you can control how much time you put into a task.
- **Input is much more certain than outcome.** When it comes to living the life you want, making sure you allocate time to living your values is the only thing you should focus on.

11

Schedule Important Relationships

LIFE DOMAINS

Family and friends help us live our values of connection, loyalty and responsibility. They need you and you need them, clearly making them far more important than a mere 'residual beneficiary', a term I first heard in an Economics 101 class. In business, a residual beneficiary is the chump who gets whatever is left over when a company is liquidated – typically, not much. In life, our loved ones deserve better, and yet, if we're not careful with how we plan our time, that's exactly what they become.

One of my most important values is to be a caring, involved and fun dad. While I aspire to live out this value, being a fully present dad is not always 'convenient'. An email from a client informs me that my website is down; the plumber texts me to say that his train has broken down and he needs to reschedule; my bank notifies me of an unexpected charge on my card. Meanwhile, my daughter sits there, waiting for me to play my next card in our game of gin rummy.

To combat this problem, I've intentionally scheduled time with my daughter every week. Much like I schedule time for a business meeting or time for myself, I block out time on my schedule to be with her.

To make sure we always have something fun to do, we spent one afternoon writing down over a hundred things to do together in town, each one on a separate little strip of paper. Then we rolled up all the little strips and placed them inside our 'Fun Jar'. Now, every Friday afternoon, we simply pull an activity from the Fun Jar and do it. Sometimes we'll visit a museum, while other times we'll play in the park or visit a highly rated ice-cream parlour across town. That time is reserved just for us.

Truth be told, the Fun Jar idea doesn't always work as smoothly as I'd like. It's hard for me to muster the energy to head for the playground when New York temperatures fall below freezing. On those days, a cup of hot cocoa and a couple of chapters of *Harry Potter* sound way more inviting for us both. What's important, though, is that I've made it a priority in my weekly schedule to live up to my values. Having this time in my schedule allows me to be the dad that I envisaged myself to be.

Similarly, my wife Julie and I make sure we have time scheduled for each other. Twice a month, we plan a special date. Sometimes we see a performance or indulge in an exotic meal. But mostly we just walk and talk for hours. Regardless of what we do, we know that this time is cemented into our schedules and will not be compromised. In the absence of this scheduled time together, it's too easy to fill our time with other errands, like running to the grocery store or cleaning the house. My time with Julie allows me to live out my value of intimacy – there's no one else I can open up to the way I can with her, but this can only happen if we make the time.

Equality is another value in my marriage. I always thought I behaved in a way that upheld that value. I was wrong. Before my wife and I had a clear schedule in place, we found ourselves bickering about why certain tasks weren't getting done around the house. Several studies show that among heterosexual couples, husbands don't do their fair share of the housework,[1] and I was, I'm sad to admit, one of them. Darcy Lockman, a psychologist in New York City, wrote in the *Washington Post*, 'Employed women partnered with employed men carry 65 percent of the family's child-care responsibilities, a figure that has held steady since the turn of the century.'[2]

But like many men Lockman interviewed in her research, I was somehow oblivious to the tasks my wife handled. As one mother told Lockman:

He's on his phone or computer while I'm running around like a crazy person getting the kids' stuff, doing the laundry.

He has his coffee in the morning reading his phone while I'm packing lunches, getting our daughter's clothes out, helping our son with his homework. He just sits there. He doesn't do it on purpose. He has no awareness of what's happening around him. I ask him about it and he gets defensive.

It was as if Lockman had interviewed my wife. But if my wife wanted help, why didn't she just ask? I later came to realise that figuring out how I could be helpful was itself work. Julie couldn't tell me how I could help because she already had a dozen things on her mind. She wanted me to take the initiative, to jump in and start helping out. But I didn't know how. I had no idea, so I'd either stand there confused or slink off to do something else. Too many evenings followed this script, ending in late dinners, hurt feelings, and sometimes tears.

During one of our date days, we sat down and listed all of the household tasks that each of us performed, making sure nothing was left out. Comparing Julie's (seemingly endless) list to mine was a wake-up call that my value of equality in our marriage needed some help. We agreed to split the household jobs and, most importantly, timeboxed the tasks on our schedules, leaving no doubt about when they would get done.

Working our way towards a more equitable split of the housework restored integrity to my value of equality in my marriage, which also improved the odds of having a long and happy relationship. Lockman's research supports this benefit: 'A growing body of research in family and clinical studies demonstrates that spousal equality promotes marital success and that inequality undermines it.'

There's no doubt scheduling time for family and ensuring they were no longer the residual beneficiary of my time greatly improved my relationship with my wife and daughter.

The people we love most should not be content with getting whatever time is left over. Everyone benefits when we hold time on our schedule to live up to our values and do our share.

This domain extends beyond just family. Not scheduling time for the important relationships in our lives is more harmful than most people realise. Recent studies have shown that a dearth of social interaction not only leads to loneliness but is also linked to a range of harmful physical effects. In fact, a lack of close friendships may be hazardous to your health.

Perhaps the most compelling evidence that friendships affect longevity comes from the ongoing Harvard Study of Adult Development.[3] Since 1938, researchers have been following the physical health as well as social habits of 724 men. Robert Waldinger, the study's current director, said in a TEDx talk, 'The clearest message that we get from this 75-year study is this: Good relationships keep us happier and healthier. Period.' Socially disconnected people are, according to Waldinger, 'less happy; their health declines earlier in midlife; their brain functioning declines sooner; [and] they live shorter lives than people who are not lonely.'[4] Waldinger warned, 'It's not just the number of friends you have ... it's the quality of your close relationships that matters.'

What makes for a quality friendship? William Rawlins, a professor of interpersonal communications at Ohio University who studies the way people interact over the course of their lives, told *The Atlantic* that satisfying friendships need three things: 'Somebody to talk to, someone to depend on, and someone to enjoy.'[5] Finding someone to talk to, depend on and enjoy often comes naturally when we're young, but as we grow into adulthood the model for how to maintain friendships is less clear. We graduate and go our separate ways, pursuing careers and starting new lives miles apart from our best friends.

Suddenly work obligations and ambitions take priority over having beers with buddies. If children enter the picture, exhilarating nights on the town become exhausted nights on the couch. Unfortunately, the less time we invest in people the easier it is to make do without them, until one day it is too awkward to reconnect.

This is how friendships die – they starve to death.

But, as the research reveals, by allowing our friendships to starve we're also malnourishing our own bodies and minds. If the food of friendship is time together, how do we make the time to ensure we're all fed?

Despite our busy schedules and surfeit of children, my friends and I have developed a social routine that ensures regular get-togethers. We call it the 'kibbutz', which in Hebrew means 'gathering'. For our gathering, four couples, my wife and me included, meet every two weeks to talk about one question – imagine an interactive TED talk over a picnic lunch. The question might range from a deep inquiry, like, 'What is one thing you are thankful your parents taught you?' to a more practical question like, 'Should we push our kids to learn things they don't want, like playing the piano?'

Having a topic helps in two ways: first, it gets us past the small talk of sport and weather, giving us an opportunity to open up about stuff that really matters. Second, it prevents the gender split that occurs when couples convene in groups – men in one corner, women in another. Having a question of the day gets us all talking together.

The most important element of the gathering is its consistency; rain or shine, the kibbutz appears on our calendars every other week – same time, same place. There's no back-and-forth emailing to hammer out logistics. To keep it even simpler, each couple brings their own food so there's no prep or clean-up. If one couple can't make it, no big deal; the kibbutz carries on as planned.

The gathering lasts about two hours, and I always leave with new ideas and insights. Most importantly, I feel closer to my friends. Given the importance of close relationships, it's essential we plan ahead. Knowing there is time set aside for the kibbutz ensures it happens.

No matter what kind of activity fulfils your need for friendship, it's essential to make time in your calendar for it. The time we spend with our friends isn't just pleasurable – it's an investment in our future health and wellbeing.

★ REMEMBER THIS:

- **The people you love deserve more than getting whatever time is left over.** If someone is important to you, make regular time for them on your calendar.
- **Go beyond scheduling date days with your significant other.** Put domestic chores on your calendar to ensure an equitable split.
- **A lack of close friendships may be hazardous to your health.** Ensure you maintain important relationships by scheduling time for regular get-togethers.

12

Sync with Stakeholders at Work

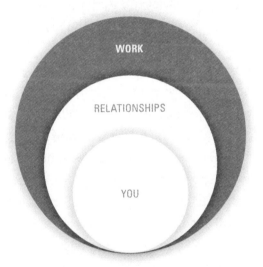

LIFE DOMAINS

WORK

RELATIONSHIPS

YOU

Unlike the other life domains, I don't need to remind you to make time for work. You probably don't have much of a choice when it comes to this. Given that work probably takes up more of your waking hours than any of the other domains, it's even more important to ensure the time spent there is consistent with your values.

Work can help people live their values of being collaborative, industrious and persistent. It also allows us to spend time on something meaningful when we labour away for someone else's benefit – such as for our customers or an important cause. Unfortunately, many of us find that our workday is a hectic mess, plagued by constant interruptions, pointless meetings and a never-ending flow of emails.

Thankfully, it doesn't have to be this way. We can do more and live better by clarifying our values and expectations with each other at work. Clarification around how we spend our time at work fosters and reinforces the central quality of a positive working relationship: trust.

Every company has its own policies. However, when it comes to how employees manage their workloads, many managers have little idea how their colleagues spend their time. Similarly, perhaps the biggest unknown to the employee is how they should spend their time, both inside and outside of work. How responsive should employees be after hours? Are they required to attend happy hours or other 'mandatory fun' events? Will managers and clients expect employees to meet last-minute deadlines? Should they let their spouses know to expect late-night outings when company execs drop into town?

These questions are significant because they directly affect our schedules and, subsequently, the time we have for the other domains in life. One survey found that 83 per cent of working professionals check their emails after work.[1] The same study notes that two-thirds of respondents take work-related devices, such as laptops or smartphones, with them on vacation. And about half the respondents said they've sent work-related emails during meals with family or friends.

Staying late at work or feeling pressured to reply to work-related messages after hours means spending less time with our family and friends or doing something for ourselves. If these demands become more than the employee bargained for, trust and loyalty can be eroded, along with one's health and relationships. The trouble is, we don't typically know the answers to these questions until we are already in a role.

There are also many unknowns from the employer's perspective. When tasks and projects take longer than originally planned and expectations aren't met, managers are left guessing why. Is the employee not capable? Is he not motivated? Is she looking for another job? How are they spending their time? In response to underperformance, managers ask employees to do more and work longer hours. But this common, knee-jerk reaction asks employees to give more than they expected, stressing the working relationship and prompting them to push back in subtle ways.

What does this pushback look like? While often unconscious of the fact, we find ourselves doing low-priority work, slacking off at our desks, chatting too much with colleagues and generally reducing productive output.

Other times, we (perhaps also unconsciously) sabotage our companies by doing pseudo-work: tasks that *look* like work but aren't in line with the company's top priorities. (Think: spending time on pet projects, corporate politicking, sending more emails, or holding more meetings than necessary.) This sort of pushback seems to increase when people work more hours.

In fact, studies have found that workers who spend more than fifty-five hours per week on the job have reduced productivity; this problem is further compounded by their making more mistakes and inflicting more useless work on their colleagues, resulting in getting even less done in more time.[2]

What's the solution to this madness?

Using a detailed, timeboxed schedule helps clarify the central trust pact between employers and employees.

Through regular review, the two parties can make informed decisions regarding whether the employee's time is spent appropriately and help them allocate time to more important tasks, both in and outside the workplace.

***

An advertising sales executive at a large tech company in Manhattan, April struggled with her schedule. Her friendly disposition had turned bitter as a result of the mounting pressure to sell more and do more in pursuit of a management role. Those pressures infected April's schedule in the form of more meetings, more unplanned conversations and more emails. Those additional tasks crowded out the time she had to focus on her priorities: caring for her customers, closing more sales and demonstrating greater results.

When I met April in her office, she looked frazzled. She had two months left to close more than a third of her annual sales quota of $15 million, and I could tell her mind was elsewhere. April feared she wouldn't meet her goal and had concluded that *she* was the problem – she just wasn't working hard enough and therefore had to do better. In her mind, *better* meant working even more hours.

Striving to be more productive was making April miserable and causing her to neglect the other domains of her life. But productivity itself wasn't her problem; she was a productive person who could squeeze a lot out of a small amount of time. Rather, the problem was her lack of a timeboxed schedule, compounded by the self-limiting belief that she, and not her management of time, was the problem. 'I'm too slow,' she told me over lunch one day. But there was nothing wrong with April. She wasn't slow, but she was lacking the productivity tools for her new role.

Though scheduling her time at work didn't come naturally to her, April subdivided her workday to account for the most important tasks she wanted to accomplish. She carved out time for focused work first, aware that creating new client proposals could be done faster and better if she did them without interruption. Every diversion slowed her down and made it more difficult to get back to customising the pitches. Then she reserved a block of time for client calls and meetings, followed by time in the afternoon for processing emails and messages. I encouraged April to share her work-related timeboxed schedule with her manager, David.

To her surprise, when they sat down to discuss her schedule, April found that David was extremely supportive of her intention to stick to a more planned-out day. 'He knew I was burning the candle at both ends,' she told me. 'When I proposed a weekly schedule, he actually seemed relieved. He told me it was helpful to know when he could call or message me instead of guessing if I was with my family.'

When she sat down with David, she realised that many of the commitments clogging her calendar weren't nearly as important to him as the time she spent closing deals. Thanks to their newfound alignment, David agreed she didn't need to attend so many meetings or mentor so many people and reassured her that this would not adversely affect her career ambitions, as long as she put in the time for her most important task: increasing revenue.

To help them stay in sync, April and David decided to meet for fifteen minutes every Monday morning at eleven o'clock. Reviewing her schedule for the week ahead would reassure them both that April was spending her time well and enable them to adjust accordingly if necessary. At the end of the meeting, she realised she could gain greater control over her workday and also cut back on the time she spent tethered to her phone at night – time that came at the expense of her personal life. April loved the outcome: a detailed view of her entire week that respected her values, reduced distractions, and, ultimately, granted her more time to do what she really wanted.

April's story is not everyone's story. The way April allocated her time won't be the way you spend your time, but schedule syncing is essential, whether with a family member or an employer. Regularly aligning expectations around how you'll spend your time is paramount, and must be done in regular, predictable increments.

If your schedule can be synced weekly, then review it and get agreement for that period, but if your schedule changes daily, getting into the routine of a brief, daily check-in with your manager will serve you both well. If you report to multiple bosses, a timeboxed calendar can serve as a way to get alignment around how you spend your time. There's no mystery about what's getting done when there's transparency in your schedule.

Remember, the Indistractable Model has four parts. Mastering internal triggers is the first step and making time for traction is the second, but there's much more we can do, as you'll soon learn. In Part 5, we'll also dive into the role of workplace culture and why persistent distraction is often a sign of organisational dysfunction. For now, it's important not to shortchange the simple yet highly effective technique of schedule syncing.

Whether at work, at home or on our own, planning ahead and timeboxing our schedules is an essential step to becoming indistractable. By defining how we spend our time and syncing with the stakeholders in our lives, we ensure that we do the things that matter and ignore the things that don't. It frees us, as it did April and me, from the trivialities of our day and gives us back the time we can't afford to waste.

But once we've reclaimed that time, how do we get the most out of it? We'll explore that question in Part 3.

★ REMEMBER THIS:

- **Syncing your schedule with stakeholders at work is critical for making time for traction in your day.** Without visibility of how you spend your time, colleagues and managers are more likely to distract you with superfluous tasks.
- **Sync as frequently as your schedule changes.** If your schedule template changes from day to day, have a daily check-in. However, most people find a weekly alignment is sufficient.

Part 3

Hack Back External Triggers

Hack Back
EXTERNAL TRIGGERS

13

Ask the Critical Question

Wendy, a freelance marketing consultant, knew exactly what she had to do for the next hour at work:[1] her calendar told her that she needed to be in her office chair at 9 a.m. to write new client proposals, the most important task of her day.

She fired up her laptop and opened the client's file on her screen, eager to win new business. As she held her coffee mug with both hands and took a sip, a fantastic addition to the proposal entered her head. 'This is going to be great!' she thought to herself.

But before she had a chance to write down the idea, '*Ping!*' her phone buzzed with a notification. At first Wendy ignored the intrusion. She jotted down a few words, but then the phone buzzed again with a different notification. This time her focus faltered, and she became curious. What if a client needed her?

She picked up her phone, only to find out that a trivial tweet by a celebrity rapper was reverberating through social media. After she tapped out of the app, another notification caught her eye. Her mother messaged her to say good morning. Wendy fired off a quick emoji heart to let her know she was fine. Oh, and what's this? A bright red notification bubble over the professional social networking app, LinkedIn. Perhaps there was a new business opportunity waiting for her? Nope; just a recruiter who had seen her profile and liked what he saw.

Wendy was tempted to reply, but she remembered the time. It was now 9.20, and she hadn't made any progress on her proposal. Worst of all, she'd forgotten the big idea she had been so excited to add to it. 'How did this happen?' she moaned to herself. Despite having important work to do, Wendy wasn't getting it done. She was, once again, distracted.

Does this sound familiar? Many of us have experienced just that kind of morning. The source of the distraction during these moments, however, isn't an internal trigger. The ubiquity of external triggers, like notifications, pings, dings, alarms and even other people, makes them hard to ignore.

It's time for us to hack back. In tech speak, 'to hack' means to 'gain unauthorised access to data in a system or computer'.[2] Similarly, our tech devices can gain unauthorised access to our brains by prompting us to distraction. Facebook's first president, Sean Parker, admitted as much when he described how the social network was designed to manipulate our behaviour.[3] 'It's a social-validation feedback loop,' he said. 'Exactly the kind of thing that a hacker like myself would come up with, because you're exploiting a vulnerability in human psychology.'

To start hacking back, we first need to understand how tech companies use external triggers to such great effect. What exactly is the 'vulnerability in human psychology' Parker described that makes us susceptible to the external triggers that so often lead to distraction?

In 2007, Dr B. J. Fogg, founder of Stanford University's Persuasive Technology Lab, taught a class on 'mass interpersonal persuasion'. Several of the students in attendance would later pursue careers applying his methods at companies like Facebook and Uber. Mike Krieger, a co-founder of Instagram, created a prototype of the app in Fogg's class that he eventually sold for $1 billion.

As a student at Stanford's Business School at the time, I attended a retreat at Fogg's home, where he taught his methods of persuasion in more depth. Learning from him firsthand was a turning point in my understanding of human behaviour: he taught me a new formula that changed the way I viewed the world.

The Fogg Behaviour Model states that for a behaviour (B) to occur, three things must be present at the same time: motivation (M), ability (A) and a trigger (T). More simply, B=MAT.

'Motivation' is 'the energy for action', according to Dr Edward Deci, professor of psychology at the University of Rochester.[4] When we're highly motivated, we have a strong desire, or energy, to take an action, and when we're not motivated we lack the energy to perform a task. Meanwhile, 'ability', in Fogg's formula, relates to how difficult or easy a behaviour is to do. The harder something is to do, the less likely people are to do it. Conversely, the easier something is to do, the more likely we are to do it.

When a person has sufficient motivation and ability, they're primed to behave in a particular way. However, without the critical third component, the behaviour will not occur. A trigger to tell us what to do next is always required.

We discussed internal triggers in a previous section, but when it comes to the products we use every day and the interruptions that lead to distraction, external triggers – stimuli in our environment that prompt us to act – play a big role.

Today, much of our struggle with distraction is a struggle with external triggers.

When BlackBerry launched push email in 2003, users rejoiced; they didn't need to constantly check their inbox for fear they'd miss important messages. 'When email comes, BlackBerry promised, your phone will tell you,' David Pierce wrote in *Wired* magazine.[5] Apple and Google soon followed and made notifications part of their phone operating systems. 'Suddenly, there was a way for anyone to jump into your phone when they wanted your attention,' Pierce continued. 'Push notifications proved to be a marketer's dream: they're functionally impossible to tell apart from a text or email without looking, so you have to look before you can dismiss.'

Checking those notifications comes at a high price. External triggers can rip us away from the task we planned to carry out at that time. Researchers have found that when people are interrupted during a task, they tend subsequently to make up for lost time by working faster, but the cost is higher levels of stress and frustration.[6]

The more we respond to external triggers, the more we train our brain in a never-ending stimulus–response loop. We condition ourselves to respond instantly. Soon, it feels impossible to do what we've planned because we're constantly reacting to external triggers instead of attending to what's in front of us.

Perhaps the answer is simply to ignore the external triggers. Maybe if we don't act on the notifications, phone calls and interruptions, we can go about our business and quickly silence the interruptions when they happen.

Not so fast. A study published in the *Journal of Experimental Psychology: Human Perception and Performance* found that receiving a cell phone notification but not replying to it was just as distracting as responding to a message or call.[7] Similarly, the authors of a study conducted at the University of Texas at Austin proposed that 'the mere presence of one's smartphone may impose a "brain drain" as limited-capacity attentional resources are recruited to inhibit automatic attention to one's phone, and are thus unavailable for engaging with the task at hand'. By having your phone in your field of view, your brain must work hard to ignore it, but if your phone isn't easily accessible or visually present, your brain is able to focus on the task at hand.

Thankfully, not all external triggers are harmful to our attention. In many ways, we can leverage them to our advantage. For example, short text messages providing words of encouragement are effective at helping smokers quit.[8] A meta-study of interventions from ten countries found that 'The evidence provides unequivocal support for the efficacy of text messaging interventions to reduce smoking behaviour', according to an author of the study.[9]

The trouble is, despite the potential benefits external triggers can provide, receiving too many can wreak havoc on our productivity and happiness. How, then, can we separate the good external triggers from the bad? The secret lies in the answer to a critical question:

Is this trigger serving me, or am I serving it?

Remember that, as the Fogg Behaviour Model describes, any behaviour requires three things: motivation, ability and a trigger. The good news is that removing unhelpful external triggers is a simple step towards controlling unwanted distractions.

When I challenged Wendy, the marketing consultant struggling to stay focused, to ask herself the critical question, it empowered her to start putting unhelpful external triggers in their place. She could begin to decide for herself which triggers led to traction instead of allowing her attention to be controlled by other people.

Viewed through the lens of this critical question, triggers can now be identified for precisely what they are: tools. If we use them properly, they can help us stay on track. If the trigger helps us do the thing we planned to do in *our* schedule, it's helping us gain traction. If it leads to distraction, then it isn't serving us.

In the next chapters, we will look at some very practical ways to manipulate our technology and our physical environment to eliminate unhelpful external triggers. We're going to hack back our devices in ways their makers never intended, but that's exactly the point – our technology should serve us, not the other way around.

⚑ REMEMBER THIS:

- **External triggers often lead to distraction.** Cues in our environment like the pings, dings, rings from devices, as well as interruptions from other people, frequently take us off-track.
- **External triggers aren't always harmful.** If an external trigger leads us to traction, it serves us.
- **We must ask: is this trigger serving me, or am I serving it?** Then, we can hack back the external triggers that don't serve us.

14

Hack Back Work Interruptions

Hospitals are supposed to help heal the sick. How, then, do we explain the 400,000 Americans harmed in hospitals every year when patients are given the wrong medication? In addition to the human toll, these preventable errors cost an estimated $3.5 billion in extra medical expenses.[1] According to Dr Martin Makary and Michael Daniel of Johns Hopkins University, 'If medical error was a disease, it would rank as the third leading cause of death in the US.'[2]

Becky Richards was part of a special team tasked with developing ways to save lives by fixing the medication-error problem at the Kaiser Permanente South San Francisco Medical Center. As a registered nurse, Richards knew many of the mistakes occurred when highly trained, well-intentioned people made very human errors that were often a result of a work environment filled with distracting external triggers. In fact, studies found nurses experienced five to ten interruptions each time they dispensed medication.[3]

One of Richards' solutions did not go over particularly well with her nursing colleagues, at least at first. She proposed nurses wear brightly coloured vests (the equivalent of hi-vis sleeveless jackets) to let others know they were dispensing medication and should not be interrupted. 'They felt it was demeaning,' Richards said in an article on nursing website RN.com.[4] After initial resistance, she found one group of nurses in an oncology unit whose error rate was particularly high and who were desperate for a solution.

However, despite these nurses' initial willingness, the test was met with more objections than Richards anticipated. For one, the orange vests looked 'cheesy', and some complained they were uncomfortably hot. They also brought interruptions from doctors who wanted to know what the vests were about. 'We were really thinking about abandoning the whole idea, because the nurses did not like it,' Richards said.

It wasn't until the hospital administration provided Richards with the results of her experiment four months later that the impact of the trial became clear. The unit recruited for Richards' experiment saw a 47 per cent drop in errors, all thanks to nothing more than wearing the vests and learning about the importance of an interruption-free environment.

'At that point we knew we could not turn our backs on our patients,' added Richards. One by one, nurses started sharing the practice, until it spread throughout the hospital and to other care centres. Some hospitals even devised their own unique solutions, like creating a specially marked 'sacred zone' on the floor where nurses prepared medications.[5] Others created special distraction-free rooms or blacked-out windows so nurses couldn't be interrupted while they worked.

More data emerged about how effective these practices were at reducing errors by shutting out unwanted external triggers.

*A multi-hospital study coordinated by the University of
California San Francisco found an 88 per cent drop in the
number of errors over a three-year period.*[6]

Julie Kliger, director of the university's Integrated Nurse
Leadership Program, told *SFGate* in 2009 that her inspiration
to expand the programme came from an unlikely place – the
airline industry. It's called the 'sterile cockpit' rule, a series
of regulations passed in the 1980s after several accidents
occurred as a result of distracted pilots. The regulations banned
commercial pilots from performing any non-critical activities
when flying under 10,000 feet. The regulation specifically calls
out 'engaging in non-essential conversations' and bars flight
attendants from contacting pilots during the most dangerous
parts of the flight: takeoffs and landings.[7]

'We liken it to flying a 747,' Kliger said. '[The zone of dangerous
distraction] for them is anything under ten thousand feet …
In the nurses' world, it's when giving medications.' Richards
reports that nurses not only make fewer mistakes while wearing
the vests, but also feel that focused work time passes more
quickly. Suzi Kim, a nurse at Kaiser West Los Angeles, said that
while wearing them, 'we can think clearly.'[8]

***

While the impact of distraction is rarely as lethal as it is for
those in the medical profession, interruptions clearly have an
impact on our work performance for any job requiring focus.
Unfortunately, interruptions are pervasive in today's workplace.

The misuse of space is often a significant contributing
factor. Seventy per cent of American offices are arranged as
open-floor plans.[9] Instead of individual workspaces separated

by walls, workers today probably have a clear line of sight to their colleagues, the break room, reception and, well, virtually everything else.

Open-office floor plans were supposed to foster idea-sharing and collaboration. Unfortunately, according to a 2016 meta-study of more than 300 papers, the trend has led to more distraction.[10] Not surprisingly, these interruptions have also been shown to decrease overall employee satisfaction.[11]

Given the toll distractions can take on our cognitive capabilities, it's time we took action, just as Becky Richards did. While I'm not advocating the wearing of bright orange 'Do Not Disturb' vests at the office, nor am I insisting on a floor-plan overhaul, I *am* suggesting a solution that is explicit and effective at deterring interruptions from co-workers.

At the end of this book, you'll find a piece of card stock. (If you're reading an e-book edition, you can download and print your own by visiting NirAndFar.com/Indistractable.) The card contains, in a large font, a simple request to passers-by, 'I need to focus right now but please come back soon'. Place the card on your computer monitor to let your colleagues know that you don't want to be interrupted. It sends an unambiguous message in a way that wearing headphones can't.

While the screen sign can be understood by just about anyone, I recommend discussing its purpose with your co-workers. This conversation could inspire them to do the same and can serve as an entry point to a discussion about the importance of working without distraction.

Sometimes, though, we need an even more explicit way to signal our request for interruption-free time, particularly when we're working from home. Using the same principles

Just as bright vests reduce prescription errors, so a screen sign sends a
signal to co-workers that you're indistractable.

to block unwanted external triggers, my wife bought a hard-
to-miss headpiece on Amazon for just a few dollars. She calls
it the 'concentration crown,' and the built-in LEDs light up
her head to send an impossible-to-ignore message. When
she wears it, she's clearly letting our daughter (and me) know
not to interrupt her unless it's an emergency. It works like
a charm.

Whether it's a vest, a screen sign or a light-up crown,
the way to reduce unwanted external triggers from other
people is to display a clear signal that you do not want to be
interrupted. Doing so will help colleagues or family members
pause and assess their own behaviours *before* they break your
concentration.

When working from home, family members can be a source of distraction. My wife's 'concentration crown' lets us know she is indistractable.

★ REMEMBER THIS:

- **Interruptions lead to mistakes.** You can't do your best work if you're frequently distracted.
- **Open-office floor plans increase distraction.**
- **Defend your focus.** Signal when you do not want to be interrupted. Use a screen sign or some other clear cue to let people know you are indistractable.

15

Hack Back Email

Email is the curse of the modern worker. Some basic maths reveals just how big the problem is. The average professional employee receives one hundred messages per day.[1] At just two minutes per email, that adds up to three hours and twenty minutes per day. If an average workday is nine to five, minus an hour for lunch, then email eats up nearly half the day.

Realistically, though, that's a very conservative estimate, since those three hours and twenty minutes don't include the time wasted in getting back to the job in hand between checking emails. In fact, a study published in the *International Journal of Information Management* found office workers took an average of sixty-four seconds after checking emails to reorient themselves and get back to work.[2] Given the hundreds of times per day we check our devices, those minutes can add up.

Lest you think email time is well spent, researchers have concluded that an astonishing number of workplace emails are an utter waste of time. 'We estimate that 25 per cent of that time is consumed reading emails that should not have been sent ... and 25 per cent is spent responding to emails that ... should never have [been] answered,' the researchers wrote in the *Harvard*

Business Review.[3] In other words, about half the time we spend on emails is as productive as counting cracks in the ceiling.

Why is email such a persistent problem? The answer can be found in understanding our psychology. Email is perhaps the mother of all habit-forming products. For one thing, it provides a variable reward. As the psychologist B. F. Skinner famously discovered, pigeons pecked at levers more often when given a reward on a variable schedule of reinforcement. Similarly, email's uncertainty keeps us checking and pecking.[4] It provides good news and bad, exciting information as well as frivolity, messages from our closest loved ones and from anonymous strangers. All that uncertainty provides a powerful draw to see what we might find when we next check our inbox. As a result, we keep clicking or pulling to refresh in a never-ending effort to quell the discomfort of anticipation.

Secondly, we have a strong tendency for reciprocity – responding in kind to the actions of another. When someone says 'Hello' or extends their hand to shake our own, we feel the urge to reciprocate – not doing so breaks a strong social norm and feels cold. Though the grace of reciprocity works well in person, it can lead to a host of problems online.

Finally, and perhaps most materially, email is a tool we have little choice but to use. For most of us, our jobs depend on it and it is so woven into our daily work and personal lives that giving it up would be a threat to our livelihood.

However, like many things in life that take more time and attention than we'd like, we can get email under control. There are techniques we can deploy as part of our working routines to defuse the unhealthy magnetism of email. Let's focus on a few techniques that deliver the best results with the least effort.

The amount of time we spend on email can be boiled down to an equation. The total time spent on email per day (T) is a function of the number of messages received (n) multiplied by the average time (t) spent on each message, so $T = n \times t$. I like to remember 'TNT' to remind me how email can blow up a well-planned day.

To reduce the total amount of time we spend on email per day, we need to address both the 'n' and 't' variables. Let's first explore ways to reduce the 'n', the total number of messages received.

***

Given our tendency for reciprocity, when we send a message it is likely the receiver will reply immediately, perpetuating the endless cycle.

To receive fewer emails, we must send fewer emails.

It seems obvious, but most of us don't observe this basic fact. So strong is our need to reciprocate that we reply to messages moments after they're received – nights, weekends, holidays, it doesn't seem to matter.

Most emails we send and receive are not urgent. Yet our brain's weakness for variable rewards makes us treat every message, regardless of form, as if it's time-sensitive. That tendency conditions us to check constantly, return replies and bark out whatever requests come to mind instantaneously. These are all mistakes.

OPEN UP 'OFFICE HOURS'

In my case, I receive dozens of emails every day asking me to discuss something related to my books or articles. I love talking with my readers, but if I responded to each email I wouldn't have time for anything else. Instead, to reduce the number of emails I send and receive, I schedule 'office hours'. Readers can book a fifteen-minute time slot with me on my website at NirAndFar.com/schedule-time-with-me.

Next time you receive a non-urgent question over email, try replying with something like, 'I've held some time on Tuesday and Thursday from 4 to 5 p.m. If this is still a concern then, please stop by and let's discuss this further.' You can even set up an online scheduling tool like mine to let people book a slot.

You'd be amazed how many things become irrelevant when you give them a little time to breathe.

By asking the other party to wait, you've given them the chance to come up with an answer for themselves – or, as is often the case, time for the problem just to disappear under the weight of some other priority.

But what if the sender still needs to discuss the question and can't figure out the problem for themselves? All the better! Difficult questions are better handled in person than over email, where there is more risk of misunderstandings. The bottom line is that asking people to discuss complex matters during regular office hours will lead to better communication and fewer emails.

SLOW DOWN AND DELAY DELIVERY

Following the maxim that the key to receiving fewer emails is sending fewer emails, it's worth considering how we can slow down the email ping-pong game by sending emails well after we write them. After all, who made the rule that every email needs to be sent as soon as you're done writing it?

Thankfully, technology can help. Instead of banging out a reply and hitting send right away, email programmes like Microsoft Office[5] and tools like Mixmax[6] for Gmail allow us to delay a message's delivery. Whenever I reply to an email, I ask myself, 'When's the latest this person needs to see this reply?'

By clicking just one extra button before sending, the email goes out of my inbox and off my plate but is held back from being delivered to the recipient until the predetermined time I selected. Thus, fewer emails sent per day results in fewer emails sent back per day.

Not only does delaying delivery allow time for the matter to be resolved by other means, it also makes it less likely I'll receive emails when I don't want them. For example, while you might enjoy clearing out your inbox on a Friday afternoon, delaying delivery until Monday prevents you from stressing out your co-workers and helps protect your weekend from relaxation-killing replies.

ELIMINATE UNWANTED MESSAGES

Finally, there's one more highly effective method for reducing inbound emails. Every day we're targeted by an endless torrent of spam, marketing emails and newsletters. Some are helpful, but most are not.

How do we stop email messages we never want to hear from again? If the email is a newsletter you signed up for in the past but no longer find useful, the best thing you can do is hit the 'unsubscribe' button at the bottom of the email. As someone who writes such a newsletter, I can tell you that we newsletter writers want you to unsubscribe if you are no longer interested. We pay email service providers per email address on our list, so prefer to send only to those who find them useful.

However, some spammy marketers make it hard to find the unsubscribe button, or might even stubbornly keep sending you emails after you've unsubscribed. For such cases, I recommend sending them into the 'black hole'. I use SaneBox, a simple programme that runs in the background as I use email.[7] Whenever I encounter an email I absolutely never want to hear from again, I click a button to send that sender's email to my SaneBlackHole folder. Once it's there, SaneBox's software ensures I'll never hear from that sender again.

Of course, managing unwanted email messages takes time, but by reducing the likelihood of unwanted messages creeping into your inbox, you'll see the number dwindle from a torrent to a trickle.

***

Now that we've covered ways to reduce the number of emails we receive (the 'n' in our equation), let's transition to the second variable: the amount of time ('t') we spend writing emails.

There's mounting evidence that processing your email in batches is much more efficient and less stress-inducing than checking it throughout the day.[8] This is because our brains take time to switch between tasks, so it's better to focus on answering emails all at once. I know what you're thinking – you can't wait all day to check email. I understand. I too need to check my inbox to make sure there's nothing truly urgent.

Checking email isn't so much the problem; it's the habitual rechecking that gets us into trouble.

Does this sound familiar? An icon tells you that you have an email, so you click and scroll through your inbox. While there, you read message after message to see if anything requires a reply. Later in the day, you open your inbox and, forgetting what was in the messages you read earlier, you reopen them once again. If you're anything like I used to be, you might open and reopen some messages an embarrassing number of times. What a waste!

PLAY TAG

We tend to believe that the most important thing about an email is its content, but that's not quite true. The most important aspect of an email, from a time management perspective, is how urgently it needs a reply. Because we forget when the sender needs a reply, we waste time rereading the message.

The solution to this mania is simple: only touch each email twice. The first time we open an email, only do one thing before closing it: Answer this question, 'When does this email require a response?' Tagging each email as either 'Today' or 'This Week' attaches the most important information to each new message, preparing it for the second (and last) time we open it. Of course, for super-urgent, email-me-right-now-type messages, go ahead and respond, while messages that don't need a response at all should be deleted or archived immediately.

Note that I'm not telling you to tag emails by topic or categories, only by when the message requires a response. Tagging emails in this way frees your mind from distraction because you know you'll reply during the time you've specifically allocated for this purpose in your timeboxed schedule.

In my case, I give my inbox a quick perusal before my morning coffee. Tagging each new email by when it requires a reply takes no more than ten minutes. It gives me peace of mind to know nothing will fall through the cracks. I can leave those messages alone and focus on my work until it's time to reply.

My daily schedule includes dedicated time for replying to emails I've tagged 'Today'. It's much quicker to respond to the urgent messages than to have to wade through all my emails to figure out which need a response by the end of the day. In addition, I reserve a three-hour timebox each week to plough through the less urgent messages I've tagged 'This Week'. Finally, at the end of my week I review my schedule to assess whether the time on my calendar for emailing was sufficient and adjust my timeboxed schedule for the week ahead.

Why not quickly type out a response when you first open a message? Taking two minutes to reply to an email on your phone doesn't sound like a big deal, until you realise that with the hundreds of messages we receive per day, those two minutes can quickly add up. Soon, two minutes turn into ten, fifteen or sixty, and you've wasted your day frantically banging out replies instead of focusing on what you really want to achieve.

Slaying the messaging monster requires a host of weapons to hack back this persistent source of distraction, but by experimenting with these proven techniques we can rein in the triggers that take us off-track.

★ REMEMBER THIS:

- **Break down the problem.** Time spent on email (T) is a function of the number of messages received (n) multiplied by the average time (t) spent per message ($T = n \times t$).
- **Reduce the number of messages received.** Schedule office hours, delay when messages are sent, and reduce the number of time-wasting messages reaching your inbox.
- **Spend less time on each message.** Label emails according to when each message needs a response. Reply to emails during a scheduled time in your calendar.

16

Hack Back Group Chat

Jason Fried says group chat is 'like being in an all-day meeting with random participants and no agenda'.[1] This is especially notable because the company Fried founded, Basecamp, makes a popular group chat app. But Fried understands it's in his company's interests to make sure his customers don't burn out. He offers several pieces of advice for teams using a group chat app, whether they use Basecamp, Slack, WhatsApp or other services.

'What we've learned is that group chat used sparingly in a few very specific situations makes a lot of sense,' Fried wrote in an online post. 'What makes a lot less sense is chat as the primary, default method of communication inside an organization. A slice, yes. The whole pie, no ... All sorts of eventual bad [things happen] when a company begins thinking one-line-at-a-time most of the time.'

Fried believes the tools we use can also change the way we feel at work, and consequently advises using group chat sparingly. 'Frazzled, exhausted, and anxious? Or calm, cool, and collected? These aren't just states of mind, they are conditions caused by the kinds of tools we use, and the kinds of behaviours those tools encourage.'

Even though the real-time nature of group chat is exactly what makes it so unique, Fried believes, '*Right now* should be the exception, not the rule.'[2] Here are four basic rules for effectively managing group chat:

Rule 1: Use It Like a Sauna

We should use group chat in the same way we use other synchronous communication channels. We wouldn't choose to participate in a conference call that lasted for a whole day, so the same goes for group chat. Fried recommends we 'Treat chat like a sauna – stay a while but then get out … it's unhealthy to stay too long.'

Alternatively, we might schedule a team meeting on group chat so that everyone is on at the same time. When used this way, it can be a great way to reduce in-person meetings.

It's telling that the CEO of a group chat company advises limiting the use of its product. And yet, many organisations that use these services encourage employees to lurk in the group chat sauna all day long. This is a corrosive practice that individuals can't always change on their own. We'll tackle dysfunctional company culture later in the book.

Rule 2: Schedule It

The single-line commentaries, GIFs and emojis commonly used in group chats create an ongoing stream of external triggers, often moving us further away from traction. To hack back, schedule time in your day to catch up on group chats, just as you would for any other task in your timeboxed calendar.

It's important to set colleagues' expectations by letting them know when you don't plan to be available. You can put them at ease by assuring them that you will contribute to the conversation during an allocated time later in the day, but until then you shouldn't feel guilty for turning on the 'Do Not Disturb' feature while doing focused work.

Rule 3: Be Picky

When it comes to group chat, be selective about who's invited into the conversation. Fried advises, 'Don't get everyone on the line. The smaller the chat, the better the chat.' Continuing the conference call metaphor, he states, 'A conference call with three people is perfect. A call with six or seven is chaotic and woefully inefficient. Group chats are no different. Be careful inviting the whole gang when you only need a few.' The key is to make sure that everyone present is able to add and extract value from being a part of the conversation.

Rule 4: Use It Selectively

Group chat is best avoided altogether when sensitive topics are being discussed. Remember that the ability to observe directly another person's mood, tone and non-verbal signals adds critical context to conversations. As Fried suggests, 'Chat should be about quick, ephemeral things', while 'Important topics need time, traction, and separation from the rest of the chatter.'

The trouble is that some people like to 'think out loud' in group chat, explaining their arguments and ideas in one-line blurbs. This rarely works, because trying to follow someone's thoughts in real time, while others comment with emojis and other potential distractions, is a poor way to understand the message.

Instead of using group chat for long arguments and hurried decisions, it's better to ask participants in the conversation to articulate their point in a document and share it *after* they've compiled their thoughts.

Ultimately, group chat is simply another communication channel, not so dissimilar from email or text messages. When used appropriately, it can have myriad benefits, but when abused or used incorrectly it can lead to a flood of unwanted external triggers. The secret lies in the answer to our critical question: are these triggers serving me, or am I serving them? We should use group chat where it helps us gain traction and weed out the external triggers that lead to distraction.

⚑ REMEMBER THIS:

- **Real-time communication channels should be used sparingly.** Time spent communicating should not come at the cost of time spent concentrating.
- **Company culture matters.** Changing group chat practices may involve questioning company norms. We'll discuss this topic in Part 5.
- **Different communication channels have different uses.** Rather than use every technology as an always-on channel, use the best tools for the job.
- **Get in and get out.** Group chat is great for replacing in-person meetings but terrible if it becomes an all-day affair.
- **Schedule time for group chat on your calendar.** Let colleagues know when you'll be in group chat and use the 'Do Not Disturb' function to let people know when you're out.
- **Be selective.** Group chat is good for some topics, people and conversations and bad for others.
- **Slow down conversations.** Ask participants who like to 'think out loud' in group chat to write down their ideas to share later.

17

Hack Back Meetings

Meetings today are full of people barely paying attention as they send emails to each other about how bored they are.[1] Part of the problem is that too often people schedule a meeting to save themselves from having to make the effort of solving a problem for themselves. To some, talking it through with colleagues feels better than working it out alone. Certainly, collaboration has its place, but meetings should not be used as a distraction from doing the hard work of thinking. How can we make meetings more worthwhile?

The primary objective of most meetings should be to gain consensus around a decision, not giving the meeting organiser a forum to hear themselves think. One of the easiest ways to prevent superfluous meetings is to demand two things of anyone who calls one. First, the meeting organiser must circulate an agenda of what problem will be discussed. No agenda, no meeting. Second, they must give their best shot at a solution in the form of a brief, written digest. The digest need not be more than a page or two discussing the problem, their reasoning and their recommendation.

These two steps require a bit more effort up front, but that's exactly the point. Requiring an agenda and a brief not only saves everyone time by getting to the answer faster, but also cuts down on unnecessary meetings by asking for a bit of effort on the part of the organiser before calling one.

But what about sharing collective wisdom and brainstorming? Those are good things, just not in meetings of more than two people. Unless the meeting is called because of an emergency or as an open forum to listen to employee concerns (which we'll discuss in Part 5), unique perspectives about a business challenge can be shared via email to the stakeholder responsible. Brainstorming can also be done before the meeting and is best done individually or in very small groups. When I taught at the Stanford Design School, I consistently saw how teams that brainstormed individually before coming together not only generated better ideas, but were also more likely to have a wider diversity of solutions as they were less likely to be overrun by the louder, more dominant members of the group.

Next, if the meeting is going to happen, we need to follow the same rules of synchronous communication discussed in the last chapter on group chat. Whether online or offline, the same rules of being selective about who attends and making sure to get in and out quickly apply.

Once we're in the meeting, there's a new problem: people on their devices instead of being fully present. Attendees check email or fiddle around on their phones during meetings despite the numerous studies showing that our brains are awful at absorbing information when we're not paying close attention.[2]

Watching others use their devices in meetings escalates an arms race of perceived productivity and paranoia – the impression that someone else is working while we're not increases our stress levels. Thinking about our own flooded inboxes reduces the meeting's effectiveness, and our lack of participation only serves to make the meeting less productive, less meaningful, less interesting and more boring.

To stay indistractable in meetings, we must rid them of nearly all screens. I've conducted countless workshops and have observed a stark difference between meetings in which tech use was permitted versus those that were device-free, and meetings without screens generated far more engaged discussion and better outcomes. In order to ensure that meeting time isn't wasted, we need to introduce new customs and rules.

If we are going to spend our time in a meeting, we must make sure that we are present, both in body and mind.

First, every conference room should have a charging station for devices, but make sure it is just out of everyone's reach. When attendees congregate prior to the meeting, they should be encouraged to silence their phone and plug in their devices so the meeting can proceed free of distractions. While there are specific exceptions to these customs based on the business, the only things attendees really need in a meeting are paper, a pen and perhaps some Post-it notes.

If slides need to be presented on-screen, designate one member of the team to present from their computer or have a dedicated laptop that stays in the meeting room. Rather than sparking the desires of others to use their devices, anyone attempting to use their phone or laptop during the meeting should get disapproving looks from you and your colleagues.

Despite the potential for increased engagement in tech-free meetings, some of us may be squeamish about the idea and may protest that we need our devices for taking notes or accessing files. But if we're honest with ourselves, we know that these are not always legitimate excuses.

Why do we *really* use our devices in meetings? Our technology gives us a way of being physically present but mentally absent; the uncomfortable truth is that we like to have our phones, tablets and laptops in meetings not for the sake of productivity, but for psychological escape. Meetings can be unbearably tense, awkward and exceedingly boring, and devices provide a way to manage our uncomfortable internal triggers.

Reducing unnecessary meetings by increasing the effort of calling one, following good rules of synchronous communication, and ensuring people are engaged in the meeting instead of on their devices, will make them much less awful.

Though the modern workplace is full of potential distractions, it is up to us to manage them by continually trying new ways to stay focused. Pick a few tactics you've learned from this section to try and ask a couple of colleagues if they're willing to give them a shot. Hacking back external triggers, whether in the office or on our devices, is an effective remedy for distraction that can help us work and live better.

★ REMEMBER THIS:

- **Make it harder to call a meeting.** To call a meeting, the organiser must circulate an agenda and briefing document.
- **Meetings are for consensus-building.** With few exceptions, creative problem-solving should occur before the meeting, individually or in very small groups.
- **Be fully present.** People use devices during meetings to escape monotony and boredom, which subsequently makes meetings even worse.
- **Have one laptop per meeting.** Devices in everyone's hands makes it more difficult to achieve the purpose of the meeting. With the exception of one laptop in the room for presenting information and taking notes, leave devices outside.

18

Hack Back Your Smartphone

It's clear that many people, myself included, are dependent on their smartphones. Whether it's to keep in contact with family, navigate around town or listen to audiobooks, this miracle device in our pockets has become indispensable. That same utility, however, also makes the smartphone a major source of potential distraction.

The good news is that being dependent is not the same thing as being addicted.[1] We can get the best out of our devices without letting them get the best of us. By hacking back our phones, we can short-circuit the external triggers that spark harmful behaviours.

Here are my four steps to hacking back your smartphone and saving yourself countless hours of mindless phone time. The best part is that implementing this plan takes less than an hour from start to finish, leaving no excuse for calling your phone distracting ever again.

Step 1: Remove

The first step to managing distraction on our phones is to remove the apps we no longer need. To do so, I had to ask myself the critical question of which external triggers on my phone were serving me and which were not. Based on my answers, I uninstalled apps that didn't align with my values. I kept apps for learning and staying healthy and removed news apps with blaring alerts and stress-inducing headlines.

I also deleted all games from my phone. I'm not saying you need to do the same, of course. Many games today, particularly those made by indie studios, are works of master craftsmanship and are no less entertaining or morally virtuous than quality books or films. But I decided that, for me, games didn't align with how I wanted to spend my time on my phone.

As a technophile, I love trying out the latest apps. However, after a few years I'd collected screen after screen of untouched apps that were now clogging up my phone. If you're anything like I used to be, you probably have a number of apps you never use. These apps take up storage space in our phone's memory and bandwidth when they update themselves. But, worst of all, these zombie apps fill our devices with visual clutter.

Step 2: Replace

Purging my unused apps was easy because saying goodbye to apps I never used didn't invoke an emotional response. However, the next step involved removing apps I loved.

For instance, I often found myself checking Facebook, Twitter or YouTube on my phone when I'd planned to spend time with my daughter. When I'd feel a tinge of boredom, I'd give a social network a quick pull-to-refresh. Unfortunately, this also pulled me out of the moment with my daughter. Abandoning these services entirely wasn't an option for me. I still wanted to use them to keep in touch with friends and watch interesting videos.

Because removing these services completely wasn't something I wanted to do, I found my solution by replacing when and where I used them. Since I'd set aside time for social media in my timeboxed schedule, there was no longer any need to have them on my phone. After a few minutes of hesitation, removing them from my phone felt like a breath of fresh air. I could breathe more easily knowing I could still access these services on my computer and at a time of my choosing, not whenever the app maker decided to ping me.

Perhaps the most unexpected behaviour replacement involving my mobile phone was changing the way I checked the time. As someone who hates being late, I used to glance at my phone throughout the day, which far too often caused me to get sucked into a notification on my phone's lock screen. When I started wearing a watch again, I noticed that I checked my phone far less frequently. A quick glance at my wrist told me what I needed to know and no more.*

The idea here is to find the best time and place to do the things you *want* to do. Just because your phone can seemingly do everything, that doesn't mean it *should*.

Step 3: Rearrange

Now that we are left only with our critical mobile apps, it's time to make our phones less cluttered and, consequently, less distracting. The aim is that nothing on our phones is able to pull us away from traction when we unlock our devices.

---

*Although I originally bought an Apple Watch for this purpose, I no longer use it. I prefer the Nokia Steel HR, which along with being a much less expensive smartwatch has the wonderful feature of always displaying the time, no wrist jerk required.

Tony Stubblebine, editor-in-chief of the popular Medium publication *Better Humans*, calls his phone's setup the 'Essential Home Screen'. Stubblebine was the sixth person to be employed at Twitter and is fully aware of the way that platform was designed with human psychology in mind.

Stubblebine recommends sorting your apps into three categories: 'Primary Tools', 'Aspirations' and 'Slot Machines'.[2] Primary Tools, he says, 'help you accomplish defined tasks that you rely on frequently: getting a ride, finding a location, adding an appointment. There should be no more than five or six.' He calls Aspirations 'the things you want to spend time doing: meditation, yoga, exercise, reading books, or listening to podcasts'. Third, Stubblebine calls Slot Machines 'the apps that you open and get lost in: email, Twitter, Facebook, Instagram, Snapchat, etc'.

'Now, rearrange your phone's home screen so that it includes only your Primary Tools and your Aspirations,' continues Stubblebine. 'Think of your home page as a group of apps that you feel you are in charge of. If the app triggers any mindless checking from you, move it to a different screen.'

Additionally, instead of swiping from screen to screen to locate an app you need, I recommend using the phone's built-in search function. This will reduce the risk of bumping into a distracting app if you begin sifting through all your phone's screens and app folders.

Step 4: Reclaim

In 2013, Apple announced that its servers had sent 7.4 trillion push notifications.[3] Unfortunately, few people do anything to avoid those external triggers. According to Adam Marchick,

A few minutes spent rearranging the apps on my phone removed external triggers I didn't need on my home screen.

CEO of mobile marketing company Kahuna, less than 15 per cent of smartphone users adjust their notification settings – meaning the remaining 85 per cent allow app makers to interrupt them whenever they'd like.[4]

It's up to us to make adjustments to suit our needs; the app makers won't do it for us. But which app notifications should we disable, and how? Now that we've whittled down the number of apps on our phones, we can adjust our notification settings. This step took me about thirty minutes but it was the most life-changing.

If you use an Apple iPhone, go to the Settings app and select the 'Notifications' option or, if you're on an Android device, find the 'Apps' section in Settings. From there, adjust each app's individual notification permissions to your preferences.

In my experience it is worth adjusting two kinds of notification permissions:

1. Sound – an audible notification is the most intrusive. Ask yourself which apps should be able to interrupt you when you are with your family or in the middle of a meeting. I only grant text messages and phone calls this privilege, though I also use an app that plays a chime every hour to help me stay on track with my schedule for the day.
2. Sight – after sound, visual triggers are the second most intrusive form of interruption. In my case, I only allow visual notifications in the form of those red circles on the corner of an app's icon and I grant this permission only to messaging services like my email app, WhatsApp, Slack and Messenger. These are not apps I use for emergencies, so I always know I can open them when I'm ready to do so.

The one hiccup with these two classifications is that some audible triggers can get through during my focused time or at night when I'm asleep. I only want those external triggers to get to me in case there's an emergency. Thankfully, my iPhone comes with two incredibly helpful 'Do Not Disturb' features (Android is rolling out similar functionality).

The first is the standard 'Do Not Disturb', which can be programmed to prevent all notifications reaching you, including calls and texts. However, when someone calls twice within three minutes or texts the word 'Urgent',[5] Apple's iOS knows to let the call or message go through.

The second feature is the 'Do Not Disturb While Driving' mode, which blocks calls and texts but also sends a message back to the sender that informs them you can't pick up the phone at the moment. You can even customise the message to let people know you are indistractable.

  Text Message

Customise an indistractable auto-reply using Apple's 'Do Not Disturb While Driving' feature.

It's worth noting that reclaiming your phone's external triggers does require a bit of maintenance. For instance, every time we install a new app, we need to adjust its notifications permission settings. The good news is that Apple iOS and Android are both planning to make the process of modifying notifications easier in upcoming updates to their respective operating systems.

***

There are many things you can do to remove the unwanted external triggers on your phones. As powerful as the app makers' tricks may be, they are no match for removing, replacing, rearranging and reclaiming the apps that don't serve you. By taking a fraction of the time you would otherwise spend getting distracted by your phone, you can customise it to eliminate unhelpful external triggers. A distraction-free mobile experience is well within your grasp. There's no reason you can't hack back.

⚑ REMEMBER THIS:

- **You can hack back the external triggers on your phone in four steps and in less than one hour.**
- **Remove:** Uninstall the apps you no longer need.
- **Replace:** Shift where and when you use potentially distracting apps, like social media and YouTube, to your desk instead of on your phone. Get a wristwatch so you don't have to look at your phone for the time.
- **Rearrange:** Move any apps that may trigger mindless checking from your phone's home screen.
- **Reclaim:** Change the notification settings for each app. Be very selective regarding which apps can send you sound and sight cues. Learn to use your phone's 'Do Not Disturb' settings.

19

Hack Back Your Desktop

By the looks of his laptop, Robbert van Els could be mistaken for a secret agent. His screen is an explosion of urgent files – a master control centre for managing clandestine operatives. The man-of-mystery persona is typified by a side-sliding BMW winding through an onslaught of Word documents and jpg files. Just looking at his desktop could raise your blood pressure.

But Robbert van Els is not a secret agent. He's a mess.

Apparently there's no correlation between the mayhem on one's computer and the adventure in one's life. Anyone can find themselves drowning in desktop clutter and this digital debris costs us time, degrades performance and kills concentration.

I first met Van Els at a conference where I presented a talk on digital distraction. At that time he had reached breaking point. He realised that if he was going to grow his business he needed to regain control of his attention. 'Less distraction [and] more time to focus,' he told me. Later, I learned that Van Els had taken my presentation to heart and gone even further. Over Facebook, he shared a screenshot of his new desktop and reported, 'I tested the new layout for a month now and the result works great!'

Robbert van Els' desktop screen, before.

Van Els discovered that a cluttered desktop doesn't just look ugly; it's also costly. For one, there are cognitive costs. A study by researchers at Princeton University found people performed poorly on cognitive tasks when objects in their field of vision were in disarray as opposed to neatly arranged.[1] The same effect applies to digital environments, according to a study published in the academic journal *Behaviour & Information Technology*.[2]

Unsurprisingly, our brains have a tougher time finding things when they are positioned in a disorganised manner, which means every errant icon, open tab or unnecessary bookmark serves as a nagging reminder of things left undone or unexplored.

With so many external triggers it's easy to mindlessly click away from the task at hand. According to Sophie Leroy at the University of Minnesota, moving from one thing to another hurts our concentration by leaving what she calls an 'attention residue'[3] that makes it harder to get back on track once we have been distracted.

Removing unnecessary external triggers from our line of sight declutters our workspace and frees the mind to concentrate on what's really important.

Today, Van Els' desktop couldn't be more pristine. He replaced the screeching sports car and hundreds of icons with simple white letters on a black background with an inspirational quote that reads, 'What we fear most is usually what we most need to do.'

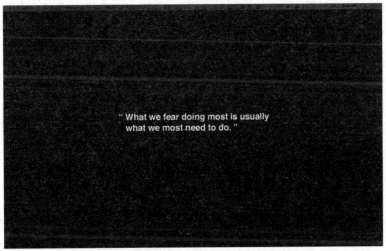
" What we fear doing most is usually what we most need to do. "

Robbert van Els' desktop today – inspiring and trigger-free.

Inspired, I decided to follow Van Els and implement a clean sweep of my own. With the exception of one or two files I will work on over the week, I put everything on my formerly cluttered desktop into one folder labelled 'Everything' (very original, I know). There's no need to sort files into folders. If I need a file,

I use the search function to find it. I now start every workday with a blank slate on my computer screen. (You can download your own Indistractable wallpaper at NirAndFar.com/Indistractable.)

But my declutter crusade didn't stop there. I decided to disable all desktop notifications to ensure that various unhelpful external triggers could no longer interrupt me. To eradicate notifications, I opened the System Preferences control panel on my Mac, clicked the 'Notifications' option and deactivated all the notification preferences for each of the listed apps.

I also hacked the 'Do Not Disturb' feature so that it remained on at all times by setting it to turn on at 7 a.m. and turn off one minute earlier. With these hacks in place, the countless desktop notifications finally stopped. Similar steps can be taken on a Windows computer using the 'Focus Assist' feature, which also includes the ability to allow interruptions from select people, like your boss.

I turned off all desktop notifications and set my laptop to perpetual 'Do Not Disturb' mode.

Like Van Els and me, you'll find that a clutter-free desktop can help you get on the path towards traction every time you switch on your computer. You will benefit from working in a digital space free of the triggers that pull your attention away from what you really want to do.

★ REMEMBER THIS:

- **Desktop clutter takes a heavy psychological toll on your attention.** Clearing away external triggers in your digital workspace can help you stay focused.
- **Turn off desktop notifications.** Disabling notifications on your computer ensures you won't get distracted by external triggers while doing focused work.

20

Hack Back Online Articles

If the internet had a voice, I'm fairly certain it would sound like HAL 9000 from *2001: A Space Odyssey*.

'Hello, Nir,' it might say to me in its low, monotone voice. 'Glad to see you again.'

'Internet, I need a few quick things for an article I'm writing,' I'd reply. 'Then it's back to work. No distractions this time.'

'Of course, Nir, but while you are here, won't you look at the news headlines?'

'No, internet,' I'd say. 'I'm just here to find some specific information. I can't be distracted.'

'Of course, Nir,' the internet would reply. 'But this article titled "The Top 10 Productivity Tricks You Just Have to Know" could be helpful. Give it a click, won't you?'

'Interesting,' I'd say hesitantly. 'Just a quick read and then it's back to work.'

Three hours later, I'd realise how long I'd wasted clicking from article to article and would curse the internet for sucking me into its content vortex yet again.

Not only was I wasting time reading too many articles, I'd often end up with dozens, if not hundreds, of open tabs strewn across my browser. These external triggers not only made me more likely to be distracted in the future, but also led to dreaded crashes, whereby all of my tabs, and whatever else I'd been working on, would be wiped out.

Thankfully, a simple rule fixed all my tab troubles and has helped me steer clear of mindless web browsing:

I never read articles in my web browser.

As you can imagine, as a writer I use the web for research every day. However, whenever I discover a new article, I no longer read it in my web browser right away. Instead, I've time-shifted *when* and *how* I read online, thereby removing the temptation to read for longer than I intend. Here's how:

I started by installing an app called Pocket on my phone, along with its browser extension on my laptop.[1] In order to abide by my 'never read articles in my browser' rule, I simply click the Pocket button in my browser every time I see an article I'd like to read. Pocket then pulls the text from the web page and saves it (without ads and any other superfluous content) to the app on my phone.

I replaced my old habit of either reading online content immediately or letting it clog up my web browser with the new habit of saving the articles for consumption at a later time. With this new behaviour, my temptation to digest the content wasn't thwarted; I was just as satisfied knowing that the content was safe and sound, waiting for me until later.

But when would I get to the hundreds of articles I'd saved? Was I merely shifting the problem from my browser to my phone? Here's where the benefits of combining timeboxing with hacking back external triggers can yield big dividends.

***

Everyone knows that multitasking destroys productivity, right? Haven't we all seen studies and read articles telling us that it's impossible to do two things at the same time? In some ways, that's true. The evidence is pretty clear that humans are awful at doing two complex tasks at once. Generally speaking, we commit more errors when juggling many tasks at the same time, and we also take longer – sometimes twice as long – to complete the tasks.[2] Scientists believe this wasted time and decreased proficiency occurs because the brain has to work hard to refocus attention.

However, when used correctly, multitasking can let us get more out of our schedules with little extra effort. I call it 'multichannel multitasking' and it's a terrific trick for getting more out of your day. To multitask the right way, we need to understand our brain's limitations that prevent us from doing more than one thing at the same time. First, the brain has a limit on its processing horsepower – the more concentration a task requires, the less room it has for anything else. That's why we can't solve two maths problems at the same time.

Second, the brain has a limited number of attention channels, and it can only make sense of one sensory signal at a time. Try listening to two different podcasts, one in each ear. Not surprisingly, you won't be able to understand what's going on in one without mentally tuning out the other.

However, although we can only receive information from one visual or auditory source at a time, we are perfectly capable of processing multichannel inputs. Scientists call this 'crossmodal attention' and it allows our brains to place certain mental processes on autopilot while we think about other things.[3]

As long as we're not required to concentrate too much on any one channel, we're able to do more than one thing at a time.

Studies have found that people can do some things better when they engage multiple sensory inputs. For example, some types of learning are enhanced when people also engage their auditory, visual and tactile senses at the same time. One study found walking, even if done slowly and on a treadmill, improved performance on a creativity test when compared to sitting down.[4]

Some forms of multichannel multitasking pair particularly well together. Cooking and eating a healthy meal with friends allows you to do something good for your body while also investing in your relationships. Stepping out of the office for a long walk while taking a phone call or inviting a colleague for a walking meeting checks off two positive things at once. Listening to a non-fiction audiobook on the way to work is a good example of making the most of a commute while investing time in self-improvement. Doing the same while cooking or cleaning makes the chores seem to pass more quickly.

Another form of multichannel multitasking has been shown to be an effective way to help people get fit. Katherine Milkman at the University of Pennsylvania's Wharton School has shown how leveraging a behaviour we 'want' to do can help us do things we know we 'should' do. In her study, Milkman gave participants an iPod loaded with an audiobook they could only listen to at the gym.[5] Milkman chose books like *The Hunger Games* and the *Twilight* series that she knew had storylines likely to keep people wanting more. The results were amazing: 'Participants who had access to the audiobooks only at the gym made 51 percent more gym visits than those in the control group.'⁶

Milkman's technique is called 'temptation bundling' and can be used whenever we want to use the rewards from one behaviour to incentivise another. In my case, the articles I save to Pocket are my rewards for exercising.

Every time I go to the gym or take a long walk, I get to listen to articles read to me through the Pocket app's text-to-speech capabilities. The built-in reading feature is astounding and the HAL 9000 voice of the internet has been replaced by a British chap with a cheery disposition who reads the articles I've selected, advert-free.

Getting through my articles feels like a small reward, often encouraging me to workout or take a stroll while satisfying my need for intellectual stimulation and saving me the temptation of reading at my desk. That, folks, is what we call a triple win in the hack back battle against distraction!

Multichannel multitasking is an underutilised tactic for getting more out of each day. We can build this technique into our schedules to help us make more time for traction and use temptation bundling to make activities like exercising more enjoyable.

My hack is one method for conquering the seductive draw of reading 'just one more thing' or having one more tab open 'for later'. By replacing my bad habits with new rules and tools, I've increased my productivity and kept HAL's seductive call at bay. Today, when online articles tempt me to keep clicking, I respond robotically, 'I'm sorry, internet, I'm afraid I can't do that.'

★ REMEMBER THIS:

- **Online articles are full of potentially distracting external triggers.** Open tabs can pull us off course and tend to suck us down a time-wasting content vortex.
- **Make a rule.** Promise yourself you'll save interesting content for later by using an app like Pocket.
- **Surprise! You can multitask.** Use multichannel multitasking like listening to articles while working out or taking walking meetings.

21

Hack Back Feeds

On the New York City subway, I often find myself surrounded by a sea of social media scrollers, their heads down as they try to reach a mythical News Feed finish line before they reach their stops. Social media is a particularly devilish source of distraction: sites like Twitter, Instagram and Reddit are designed to spawn external triggers – news, updates and notifications galore.

The infinite scroll of Facebook's News Feed is an ingenious bit of behavioural design and is the company's response to the human penchant for perpetually searching for novelty. But just because Facebook uses sophisticated algorithms to keep us tapping doesn't mean we can't hack back. I've found the most effective way to regain control is to eliminate the News Feed altogether. Didn't think that was possible? It is, and here's how.

A free web browser extension called News Feed Eradicator for Facebook does exactly what it says: it eliminates the source of countless alluring external triggers and replaces them with an inspirational quote.[1] If that tool doesn't take your fancy, another free technology called Todobook replaces the Facebook News Feed with the user's to-do list. Instead of scrolling the feed, we see tasks that we planned to do for the day, and only when we've completed our to-do list does the News Feed unlock.[2] Ian McCrystal, Todobook's founder, told *Mashable*, 'I love News Feed, I just want a more healthy relationship with it ... So I wanted a way to keep up my productivity while still having access to the less-distracting parts of Facebook.'

You can hack back Facebook by removing the News Feed.

Personally, I still use Facebook, but now I use it the way I want instead of the way Facebook intended. When I want to see updates from a certain friend or participate in the discussion happening in a particular Facbook group, I go straight to the page I want instead of having to wrestle myself away from the News Feed. I allocate time on my calendar to check Facebook almost every day, but without the unwanted external triggers in the News Feed to tempt me down a rabbit hole of frivolity; I'm in and out in less than fifteen minutes.

Though technologies like Todobook work across several other social media sites including Reddit and Twitter, there's also another way to avoid distractions on these and other feed-based social networking sites: bypass the feed using a clever bookmarking protocol.

For example, typing in 'LinkedIn.com' takes you to the website's feed, where a stream of stories can keep you scrolling and clicking for hours. While I could install a browser extension called Newsfeed Burner,[3] which eliminates the LinkedIn feed, I benefit from the industry information in the LinkedIn feed and don't want it gone completely. In this case, instead of eradicating the feed, I simply take charge of the exact URL when I visit the site, making sure I choose a destination with fewer external triggers likely to distract me.

Here's how it works: during my scheduled social media time, I click on a button in my browser to activate an extension called Open Multiple Websites.[4] As the name suggests, the button opens all the website addresses I've preloaded. Since I don't want to land on the LinkedIn.com feed, I've preloaded LinkedIn.com/messaging, where I can read and respond to messages instead of falling victim to the endless, distracting feed. With the same click, the browser extension opens Twitter.com/NirEyal, where I can respond to comments and questions without seeing the infamous and inflammatory Twitter feed.

By avoiding the feed, I'm much more likely to use social media mindfully while still allowing time to connect with others proactively.

Just as companies like Facebook and LinkedIn implement behavioural design to keep us scrolling, YouTube deploys similar psychological hacks to keep us watching with its powerful external triggers. As you watch a video, YouTube's algorithm hums away at predicting what you'll likely want to watch next, based on the topic of the video you're currently watching and your video history.[5] YouTube serves up thumbnail images of recommended videos along the right side of the web page, usually next to advertisements for sponsored videos targeted at you. Similar to a news feed, these thumbnails also appear as soon as you land on their homepage, sending you on a hunt for more digital treasure. Such external triggers are there to keep you watching video after video.

Of course, there's nothing inherently wrong with spending time on YouTube. I have time reserved in my timeboxed calendar to indulge in YouTube videos and I love it! But, rather than mindlessly viewing the next recommended video or clicking on yet another enticing suggestion, I use some hacks of my own to make sure I only watch videos I'd planned to see.

Specifically, I like the free browser extension called DF Tube, which scrubs away many of the distracting external triggers and lets me watch a video in peace.[6] I find that removing the suggested videos and ads along the side of the screen is a huge help.

Overcoming the countless external triggers on social media, from news feeds to suggested videos, represents a significant

You can hack back YouTube by removing distracting video thumbnails and ads.

step in our quest to become indistractable. Regardless of the exact tool we choose, the key is to regain control over our experiences rather than allowing feeds to control us.

★ REMEMBER THIS:

- **Feeds like the ones we scroll through on social media are designed to keep you engaged.** Feeds are full of external triggers that can drive us to distraction.
- **Take control of feeds by hacking back.** Use free browser extensions like News Feed Eradicator for Facebook, Newsfeed Burner, Open Multiple Websites, and DF Tube to remove distracting external triggers. (Links to all these services and more are available at NirAndFar.com/ Indistractable.)

Part 4

Prevent Distraction with Pacts

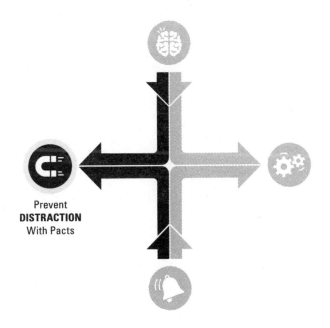

22

The Power of Precommitments

Jonathan Franzen, the writer *Time* magazine called the 'Great American Novelist', struggles with distraction just like you and me. The difference, however, between Franzen and most people is that he takes drastic steps to keep himself focused. According to a 2010 *Time* profile:

> He uses a heavy, obsolete Dell laptop from which he has scoured any trace of hearts and solitaire, down to the level of the operating system. Because Franzen believes you can't write serious fiction on a computer that's connected to the internet, he not only removed the Dell's wireless card but also permanently blocked its Ethernet port. 'What you have to do,' he explains, 'is you plug in an Ethernet cable with superglue and then you saw off the little head of it.'[1]

Franzen's methods may seem extreme, but desperate times call for desperate measures. And Franzen is not alone in his methods. Famed director Quentin Tarantino 'never use[s] a typewriter or computer', preferring to write his work by hand in a notebook.² Pulitzer Prize-winning author Jhumpa Lahiri writes her books with pen and paper and then types them on a computer without internet.³

What these creative professionals understand is that focus not only requires keeping distraction out, it also necessitates keeping ourselves in. After we've learned to master internal triggers, make time for traction and hack back external triggers, the last step to becoming indistractable involves preventing ourselves from sliding into distraction. To do so, we must learn a powerful technique called a 'precommitment', which involves removing a future choice in order to overcome our impulsiveness.⁴

Although researchers are still studying why it is so effective, precommitment is in fact an age-old tactic. Perhaps the most iconic precommitment in history appears in the ancient telling of the *Odyssey*. In the story, Ulysses must sail his ship and crew past the land of the Sirens, who sing a bewitching song known to draw sailors to their shores. When sailors approach, they wreck their ships on the Sirens' rocky coast and perish.

Knowing the danger ahead, Ulysses hatches a clever plan to avoid this fate. He orders his men to fill their ears with beeswax so they cannot hear the Sirens' call. Everyone follows Ulysses' orders, with the exception of Ulysses who wants to hear the beautiful song for himself.

But Ulysses knows that he will be tempted either to steer his ship towards the rocks or jump into the sea to reach the Sirens. To safeguard himself and his men, he instructs his crew to tie him to the mast of the ship and orders them neither to set him free, nor change course until the ship is in the clear, no matter what he says or does. The crew follows Ulysses' commands, and, as the ship passes the Sirens' shores, he is driven temporarily insane by their song. In a fury, he calls for his men to let him go, but since they cannot hear the Sirens or their captain they navigate past the danger safely.

In Homer's *Odyssey*, Ulysses resists the Sirens' song by making a precommitment and successfully avoiding the distraction.[5]

A 'Ulysses pact' is defined as 'a freely made decision that is designed and intended to bind oneself in the future,'[6] and is a type of precommitment we still use today. For example, we precommit to advanced healthcare directives to let our doctors and family members know our intentions should we lose our ability to make sound judgements. We precommit to our financial security by depositing money in retirement accounts with steep penalties for early withdrawal to ensure we don't spend funds we'll need later in life. We covet the fidelity that is promised in a lifelong relationship bound by the contract of marriage.

Such precommitments are powerful because they cement our intentions when we're clearheaded and make us less likely to act against our best interests later. Just as we make precommitments in other areas of our lives, we can utilise them in our counter-offensive against distraction.

The most effective time to introduce a precommitment is after we've addressed the first three aspects of the Indistractable Model.

If we haven't fundamentally dealt with the internal triggers driving us towards distraction, as we learned in Part 1, we'll be set up for failure. Similarly, if we haven't set aside time for traction, as we learned in Part 2, our precommitments will be useless. And finally, if we don't first remove the external triggers that don't serve us before we make a precommitment, it's probably not going to work. Precommitments are the last line of defence preventing us from sliding into distraction. In the next few chapters, we'll explore the three kinds of precommitments we can use to keep ourselves on track.

⚑ REMEMBER THIS:

- **Being indistractable not only requires keeping distraction out.** It also necessitates reining ourselves in.
- **Precommitments can reduce the likelihood of distraction.** They help us stick to decisions we've made in advance.
- **Precommitments should only be used after the other three indistractable strategies have already been applied.** Don't skip the first three steps.

23

Prevent Distraction with Effort Pacts

Inventors David Krippendorf and Ryan Tseng came up with a simple way to stop their unwanted habit of indulging in late-night snacking. Their device, Kitchen Safe, is a plastic container equipped with a locking timer built into the lid.

Placing your tempting treats (like Oreo cookies, my personal favourite) in the container and setting the Kitchen Safe timer locks the container until the timer runs out. Of course, you could smash the container with a hammer or run out to buy some more cookies, but the effort required makes that less likely to happen. Krippendorf and Tseng's concept was so compelling that it scored a deal on the reality show *Shark Tank* and the product now has nearly 400 five-star reviews on Amazon.[1]

Kitchen Safe is an example of a precommitment. Specifically, it demonstrates the usefulness of an effort pact: a kind of precommitment that involves increasing the amount of effort required to perform an undesirable action. This type of precommitment can help us become indistractable.

An effort pact prevents distraction by making unwanted behaviours more difficult to do.

An explosion of new products and services is vying to help us make effort pacts with our digital devices. Whenever I write on my laptop, for instance, I click on the SelfControl app, which blocks my access to a host of distracting websites like Facebook and Reddit, as well as my email account.[2] I can set it to block these sites for as much time as I need, typically in forty-five-minute to one-hour increments. Another app called Freedom is a bit more sophisticated and blocks potential distractions not only on my computer but also on mobile devices.[3]

Forest, perhaps my favourite distraction-proofing app, is one I find myself using nearly every day.[4] Every time I want to make an effort pact with myself to avoid getting distracted on my phone, I open the Forest app and set my desired length of phone-free time. As soon as I hit a button marked 'Plant', a tiny seedling appears on the screen and a timer starts counting down. If I attempt to switch tasks on my phone before the timer runs out, my virtual tree dies. The thought of killing the little virtual tree adds just enough extra effort to discourage me from tapping out of the app – a visible reminder of the pact I've made with myself.

Apple and Google are also joining the crusade against digital distractions by adding effort pact capabilities to their operating systems. Apple's iOS 12 allows users to schedule time constraints for certain apps through its 'Downtime' function.[5] If users attempt to access a listed app during hours that they specify, the phone prompts the user to take an additional step in order to confirm that they want to break their pact to access it. Newer versions of Google's Android come with 'Digital Wellbeing' features that provide similar functionality.

The Forest app is a simple way to make an effort pact on your phone.

Adding a bit of additional effort forces us to ask if a distraction is worth it. Whether with the help of a product like Kitchen Safe or an app like Forest, effort pacts are not limited to those we make with ourselves; another highly effective way to forge them involves making pacts with other people.

In previous generations, social pressure helped us stay on task – before the invention of the personal computer, procrastinating at our desks was obvious to the entire office. Reading a copy of *Sports Illustrated* or *Vogue*, or recapping the details of our long weekend while on the phone with a friend, sent clear signals to our colleagues that we were slacking off.

In contrast, few people today can see what we're scrolling through or clicking on while at the office. Hunched over our laptops, we find ourselves checking sports scores, news feeds or celebrity gossip headlines throughout the workday. To a passer-by, all this looks just the same as doing some research or following up on sales leads. Disguised by the privacy of our screens, the social pressure to stay on task disappears.

The problem becomes more acute when we work remotely. Since I tend to work from home, I find it all too easy to get off-track when I know I should be writing. Perhaps reintroducing a bit of social pressure when I'm having trouble staying focused could be helpful? I put the question to the test and asked my friend Taylor, a fellow author, to co-work with me. Most mornings we sat at adjacent desks in my home office and agreed to work in timed sprints of forty-five minutes. Seeing him hard at work, particularly at times when I found myself losing steam, and knowing that he could see me, kept me doing the work I knew I needed to do. Scheduling time with a friend for focused work proved to be an effective way to commit to doing what mattered most.

But what if you can't find a colleague with a compatible schedule? When Taylor went away for a week to speak at a conference, I needed to recreate the experience of making an effort pact with another person. Thankfully, I found Focusmate. With a vision to help people around the world stay focused, they facilitate effort pacts via a one-to-one video conferencing service.

While Taylor was away, I signed up at FocusMate.com and was paired with a Czech medical school student named Martin. Because I knew he would be waiting for me to co-work at our scheduled time, I didn't want to let him down. While Martin was hard at work memorising human anatomy, I stayed focused on my writing. To discourage people from skipping their meeting times, participants are encouraged to leave a review of their focus mate.*

Effort pacts make us less likely to abandon the task at hand. Whether we make them with friends and colleagues, or via tools like Forest, SelfControl, FocusMate or Kitchen Safe, effort pacts are a simple yet highly effective way to keep us from getting distracted.

⚑ REMEMBER THIS:

- **An effort pact prevents distraction by making unwanted behaviours more difficult to do.**
- **In the age of the personal computer, social pressure to stay on task has largely disappeared.** No one can see what you're working on, so it's easier to slack off. Working next to a colleague or friend for a set period of time can be a highly effective effort pact.
- **You can use tech to stay off tech.** Apps like SelfControl and Forest can help you make effort pacts with yourself.

---

*I liked the service so much I decided to invest in FocusMate.

24

Prevent Distraction with Price Pacts

A price pact is a type of precommitment that involves putting money on the line to encourage us to do what we say we will. Stick to your intended behaviour and keep the cash; get distracted and you forfeit the funds. It sounds harsh, but the results are stunning.

A study published in the *New England Journal of Medicine* illustrated the power of price pacts by examining three groups of smokers who were trying to quit their unhealthy habit.[1] In the study, a control group was offered educational information and traditional methods, such as free nicotine patches, to encourage smoking cessation. After six months, 6 per cent of people in the control group had stopped smoking. The next group, called the 'reward group', was offered $800 if they had stopped smoking after six months – 17 per cent of them were successful.

However, the third group of participants provided the most interesting results. In this group, called the 'deposit group', participants were required to make a precommitment deposit of $150 of their own money with a pledge to be smoking-free after six months. If, and only if, they reached their goal, they would receive the $150 deposit back. In addition to recouping their cash, successful 'deposit' group participants would also receive a $650 bonus prize (as opposed to the $800 offered to the 'reward' participants) from their employer.

The results? Of those who accepted the deposit challenge, an astounding 52 per cent succeeded in meeting their goal! One would imagine that a greater reward ought to lead to greater motivation to succeed, so why would winning the $800 reward be *less* effective than winning the $650 reward plus $150 deposit? Perhaps participants in the 'deposit' group were more motivated to quit smoking in the first place? To combat this potential bias, the study's authors only used data from smokers willing to be in either test group.

Explaining the results, one of the study's authors wrote, 'people are typically more motivated to avoid losses than to seek gains'; losing hurts more than winning feels good. This irrational tendency, known as 'loss aversion', is a cornerstone of behavioural economics.

I've learned how to harness the power of loss aversion in a positive way. A few years ago I was frustrated at the number of excuses I was making for not exercising regularly. At the time, going to the gym couldn't have been easier – the fully equipped facility was located in my apartment complex. I couldn't blame my no-shows on traffic, nor could I blame it on membership dues, because membership was free for residents. Even taking a long walk would be better than doing nothing. Yet I somehow found reasons to skip my workouts.

I decided to make a price pact with myself. After making time in my timeboxed schedule, I taped a crisp hundred-dollar bill to the calendar on my wall, next to the date of my upcoming workout. Then I bought a 99 cent lighter and placed it nearby.

Every day, I had a daily choice to make: I would either burn the calories by exercising or burn the hundred-dollar bill. Unless I was certifiably sick, those were the only two options I allowed myself.

Any time I found myself coming up with petty excuses, I had a crystal-clear external trigger that reminded me of the precommitment had I made to myself and to my health. I know what you're thinking, 'That's too extreme! You can't burn money like that!' That's exactly my point. I've used this 'burn or burn' technique for over three years and have gained twelve pounds of muscle, without *ever* burning the hundred dollars.

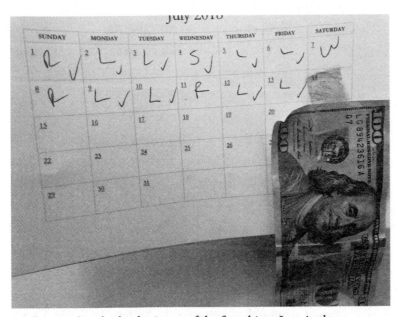

My 'burn or burn' calendar is one of the first things I see in the morning. It reminds me that I need either to burn calories or burn the hundred-dollar bill.*

'If you're curious, 'R' stands for 'run', 'L' means 'lift' (as in lift weights), 'S' stands for 'sprints', 'W' means 'walk' and the check mark indicates I did my writing for the day.

As exemplified by my 'burn or burn' method, a price pact binds us to action by attaching a price to distraction. But a price pact need not be limited to smoking cessation, weight loss or fitness goals; in fact, I found it helpful for achieving my professional ambitions as well. After spending nearly five years conducting the research for this book, I knew it was finally time to start putting words on the page, but found it difficult to get down to writing each day and instead found myself doing even more research, both online and offline. Even worse, I found myself a few clicks away from consuming media that was entirely irrelevant to my writing goals. Clearly, I was not making traction.

Eventually, I'd had enough with my false starts, half-finished chapters and incomplete outlines. I decided to put some skin in the game and enter a price pact to hold myself accountable to my important goal of finishing this book.

I asked my friend Mark to be my accountability partner in my price pact; if I didn't finish a first draft of this book by a set date, I had to pay him $10,000. The thought of it made me sick to my stomach – if I forfeited the money, gone would be the vacation budget I'd set aside for my fortieth birthday; gone would be my self-indulgent fund reserved for my new adjustable desk; most devastatingly, gone would be the completion of this book, a goal I so desperately wanted to achieve.

A price pact is effective because it moves the pain of losing to the present moment as opposed to a distant future. There's also nothing special about the dollar amount so long as it hurts to lose the sum. For me, the price pact worked like a charm, because knowing that I had so much on the line kicked me into high gear. I committed to a minimum of two hours of distraction-free writing time six days per week, added it to my timeboxed schedule and got down to work each day. In the end, I was able to keep my money (and my vacation and adjustable desk) and you're now reading the result of my work.

***

By this point, you may think price pacts are an impenetrable defence against distraction. Why not just make the cost of distraction so high that you always stay on track? The fact is, price pacts aren't for everyone and for every situation. While price pacts can be highly effective, they come with some caveats. To experience the best results with price pacts, we need to be aware of and plan for their pitfalls:

Pitfall #1: Price pacts aren't good at changing behaviours with external triggers you can't escape.

There are certain behaviours that aren't suitable for changing through a price pact. This kind of precommitment is not recommended when you can't remove the external trigger associated with the behaviour.

For example, nail biting is a devilishly hard habit to break because nail biters are constantly tempted whenever they become aware of their hands. Such body-focused repetitive behaviours are not good candidates for price pacts. Similarly, attempting to finish a big project that requires intense focus while working next to a colleague who continually wants to show you the latest photos of their 'super-cute' puppy is unreasonable. Price pacts only work when you can tune out or turn off the external triggers.

Pitfall #2: Price pacts should only be used for short tasks.

Implementing price pacts like my 'burn or burn' technique work well because they require short bursts of motivation – a quick trip to the gym, two hours of focused writing time or 'surfing the urge' of a cigarette craving, for example. If we are bound by a pact for too long, we begin to associate it with punishment, which can spawn counterproductive effects, such as resentment of the task or goal.

Pitfall #3: Entering a price pact is scary.

Despite knowing how effective they are, most people cringe at the idea of making a price pact in their own lives – I sure did at first! I struggled with committing to my 'burn or burn' regimen because I knew it meant I would have to do the uncomfortable work of hitting the gym. Similarly, shaking Mark's hand and pledging to finish my manuscript made me sweat. Only later did I realise how illogical it was to resist a goal-setting technique that makes success so much more likely.

Expect a certain amount of trepidation when entering into a price pact, but do it anyway.

Pitfall #4: Price pacts aren't for people who beat themselves up.

Despite taking part in one of the most successful smoking cessation studies ever conducted, some 48 per cent of the participants in the 'deposit' group did not achieve their goal. Behaviour change is hard, and some people will fail. Any programme for long-term behaviour modification must accommodate those of us who, for one reason or another, don't stick with it. It's critical to know how to bounce back from failure. As we learned in Chapter 8, responding to setbacks with self-compassion instead of self-criticism is the way to get back on track. While trying a price pact, make sure you are able to be kind to yourself and understand that you can always adjust the programme to give it another go.

None of the four pitfalls negates the benefits of making a price pact. Rather, they are preconditions to ensuring we use the right tool for the job. When used in the right way, price pacts can be a highly effective way to stay focused on a difficult task by assigning a cost to distraction.

★ REMEMBER THIS:

- **A price pact adds a cost to getting distracted.** It has been shown to be a highly effective motivator.
- **Price pacts are most effective when you can remove the external triggers that lead to distraction.**
- **Price pacts work best when the distraction is temporary.**
- **Price pacts can be difficult to start.** We fear making a price pact because we know we'll actually have to do the thing we're scared to do.
- **Learn self-compassion before making a price pact.**

25

Prevent Distraction with Identity Pacts

One of the most effective ways to change our behaviour is to change our identity. No, this doesn't require joining a witness protection programme or the CIA. Rather, as modern psychology confirms, slight alterations in the way we see ourselves can have a dramatic effect on our future actions.

Consider an experiment run by a group of Stanford University psychologists in 2011.[1] A young researcher named Christopher Bryan designed a study to test the effects of priming individuals to think of themselves in slightly different ways. First, he asked two groups of registered voters to complete questions related to an upcoming election. One group's survey questions included the verb 'to vote'; for example, 'How important is it to you to vote?' The second group answered similar questions that included the noun 'voter'; such as, 'How important is it to you to be a voter?'[2] The difference in wording may seem minor, but the results were extraordinary.

To measure the effect of the small wording change, the researchers then asked participants of their intentions to vote and cross-referenced public voting records to confirm whether they had actually followed through. The results were 'among the largest experimental effects ever observed on objectively measured voter turnout', Bryan and his co-authors wrote in a study published by the *Proceedings of the National Academy of Sciences*.[3] They found that those shown the survey about being a 'voter' were much more likely to vote than those who were asked how likely they were 'to vote'.

The results were so surprising that the researchers replicated the experiment during another election to confirm their validity. The results were the same: the 'voter' group dramatically outperformed the 'to vote' group.

Bryan concluded, 'People may be more likely to vote when voting is represented as an expression of self – as symbolic of a person's fundamental character – rather than as simply a behaviour.'

Our self-image has a sizeable impact on our behaviour and has implications well beyond the voting booth. Identity is another cognitive shortcut that helps our brains make otherwise difficult choices in advance, thereby streamlining decision-making.

Our perception of who we are changes what we do.

The way we think of ourselves also has a profound impact on how we deal with distractions and unintended behaviours. A study published in the *Journal of Consumer Research* tested the words people use when faced with temptation.[4] During the experiment, one group was instructed to use the words 'I can't' when considering unhealthy food choices, while the other group used 'I don't'. At the end of the study, participants were offered either a chocolate bar or granola bar to thank them for their time. Nearly twice as many people in the 'I don't' group picked the healthier option on their way out of the door.

The authors of the study attributed the difference to the 'psychological empowerment' that comes with saying 'I don't' rather than 'I can't'. The results were similar to those in the voting study: 'I can't' relates to the behaviour, while 'I don't' says something about the person.

To leverage the power of identity to prevent distraction, we can enter into what I call an 'identity pact', which is a precommitment to a self-image that helps us pursue what we really want.

There's an old joke that goes, 'How do you know someone is a vegetarian?' The punchline: 'Don't worry, they'll tell you.' You could replace 'vegetarian' with any number of monikers, from marathon runner to marine, and the joke would still ring true.

I was a vegetarian for five years. As anyone who has tried a meat-free diet knows, friends always ask, 'Don't you miss meat? I mean, it tastes so good!' Of course I missed meat! However, when I began calling myself a vegetarian, somehow what was once appetising suddenly became something else. The things I once loved to eat were now unpalatable because I had changed how I defined myself. It wasn't that I *couldn't* eat meat; I was a vegetarian, and vegetarians *don't* eat meat.

When I made this identity pact, I was limiting my future choices, but saying no to meat was no longer difficult. Rather than being a chore or a burden, it became something I simply did not do, much in the same way that observant Muslims do not drink alcohol and devout Jews do not eat pork – they just don't.

By aligning our behaviours to our identity, we make choices based on who we believe we are.

With that in mind, what identity should we take on to help fight distraction? It should now be clear why this book is titled *Indistractable*. Welcome to your new moniker! By thinking of yourself as indistractable, you empower yourself through your new identity. You can also use this identity as a rationale to tell others why you do 'strange' things like meticulously plan your time, refuse to respond to every notification immediately or put a sign on your screen when you don't want to be disturbed. These acts are no more unusual than other identities people adhere to, like wearing religious garb or following a particular diet. It's time to be indistractable and proud!

Telling others about your new identity is a great way to solidify your pact. Have you noticed how many religions encourage adherents to evangelise their faith? Missionary work is a way to grow the number of adherents, but, psychologically speaking, there's more to proselytising than getting non-believers to join the fold.

According to several recent studies, preaching to others can greatly impact the motivation and adherence of the teacher. Drs Lauren Eskreis-Winkler and Ayelet Fishbach have run experiments on diverse groups, from unemployed workers looking for a job to children struggling in school. Their results consistently show that teaching others provides more motivation for the teacher to change their *own* behaviour than if the teacher learned from an expert.[5]

But do we have a right to teach others about something we haven't quite figured out ourselves? Should we preach when we're far from perfect? Studies show teaching others can be even more effective at changing our future behaviour when we admit our own struggles.[6] As Eskreis-Winkler and Fishbach note in the *MIT Sloan Management Review*,[7] when people confess past mistakes they are able to acknowledge where they've gone wrong without developing a negative self-image. Rather, teaching empowers us to construct a different identity, as shown by the act of helping other people avoid making the same mistakes.

Another way to reinforce our identity is through rituals. Let's look again at religion. Many religious practices aren't easy, at least not for outsiders. Praying five times a day towards Mecca or reciting prescribed blessings before each meal is an effort. And yet, for strict adherents, these routines are something they just do, without fail and without question. What if we could tap into that same devotion to accomplish difficult tasks? Imagine having the fortitude to focus on whatever you wanted with the same loyalty of a true believer.

New research suggests that secular rituals, in the workplace and in everyday life, can have a powerful effect. A study conducted by Professor Francesca Gino and her colleagues at Harvard Business School explored how rituals affect self-control by studying people trying to lose weight.[8] The first group was asked to be mindful of what they ate for five days, while the second group was taught a three-step pre-meal ritual. First, they had to cut their food; second, arrange the pieces symmetrically on the plate; and third, tap their food three times with their utensils before eating. Silly, yes, but also surprisingly effective. The study participants who followed the pre-eating ritual ate, on average, fewer calories, less fat and less sugar than those in the 'mindful group.'*

---

*Eating rituals aren't for everyone. People suffering from an eating disorder should not engage in them.

Professor Gino believes rituals 'may seem like a waste of time. Yet, as our research suggests, they are quite powerful.' She continues, 'Even when they are not embedded in years of tradition, simple rituals can help us build personal discipline and self-control.'

Though conventional wisdom says our beliefs shape our behaviours, the opposite is also true.[9]

Evidence of the importance of rituals supports the idea of keeping a regular schedule, as described in Part 2. The more we stick to our plans, the more we reinforce our identity. We can also incorporate other rituals into our lives to help remind us of our identity. For example, I have a ritual of repeating a series of short mantras every morning. I've collected them over the years and say them before I start my work every day. A quick reading of these snippets of indistractable wisdom, such as the William James quote 'The art of being wise is the art of knowing what to overlook', reinforces my identity through ritual.[10]

I also find opportunities to label myself as indistractable. For instance, when I'm working from home, I tell my wife and daughter that I'm indistractable prior to starting a focused work block. As you learned in Chapter 18, I use my phone's 'Do Not Disturb' function to send an auto-reply message stating that I'm indistractable to anyone who might contact me during my focused time. I even have t-shirts printed with the word 'INDISTRACTABLE' across the chest to reinforce my identity whenever I look in the mirror or someone asks me about my shirt.

By making identity pacts, we are able to build the self-image we want. Whether the behaviour is related to what we eat, how we treat others or how we manage distraction, this technique can help shape our behaviour to reflect our values. Though we often assume our identity is fixed, our self-image is, in fact, flexible and is nothing more than a construct in our minds. It's a habit of thought, and, as we've learned, habits can be changed for the better.

***

Now that you know the four parts of the Indistractable Model, you're ready to put these strategies to work. Make sure you can draw out the four parts of the model (traction/distraction, internal triggers/external triggers) so you can share the model with others as well as have ready access to it the next time you find yourself struggling with distraction.

Up to now we've focused primarily on what *you* can do to become indistractable. But we must acknowledge that we work and live with other people. In the next section, we'll dive into how workplace culture affects distraction. Then, we'll learn about why children overuse their distractions and what we can all learn from their need for 'psychological nutrients'. Finally, we'll dive into how we can be indistractable around friends and loved ones, and help them stay focused as well.

⚑ REMEMBER THIS:

- **Identity greatly influences our behaviour.** People tend to align their actions with how they see themselves.
- **An identity pact is a precommitment to a self-image.** You can prevent distraction by acting in line with your identity.
- **Become a noun.** By assigning yourself a moniker, you increase the likelihood of following through with behaviours consistent with what you call yourself. Call yourself 'indistractable'.
- **Share with others.** Teaching others solidifies your commitment, even if you're still struggling. A great way to be indistractable is to tell friends about what you learned in this book and the changes you're making in your life.
- **Adopt rituals.** Repeating mantras, keeping a timeboxed schedule, or performing other routines, reinforces your identity and influences your future actions.

Part 5

How to Make Your Workplace Indistractable

26

Distraction Is a Sign of Dysfunction

The modern workplace is a constant source of distraction. We plan to work on a big project that demands our undivided attention, but we are distracted from it by a request from our boss. We book an hour of focused work, only to be pulled into yet another 'urgent' meeting. We might make time to be with our family or friends after hours, only to be called into a late-night video conference call.

Though we've discussed various tactics in earlier chapters, including timeboxing, schedule syncing and hacking back external triggers in the workplace, for some of us the problem is bigger than upgrading our skills.

While learning to control distractions on our own is important, what do we do when our jobs repeatedly insist on interrupting our plans? How can we do what is best for our careers, not to mention our companies, when we're constantly distracted? Is today's always-on work environment the inescapable new normal or is there a better way?

To many, the adoption of various technologies appears to be the source of the problem. After all, as technologies like email, smartphones and group chat proliferated through enterprises, employees were expected to use these tools to deliver whatever their managers wanted, whenever they wanted it. However, new research into why we get distracted at work reveals a deeper cause.

As we learned in Part 1, distractions originate from a need to escape psychological discomfort. So what is making the modern employee so uncomfortable? There is mounting evidence that some organisations make their employees feel a great deal of pain. In fact, a 2006 meta-analysis by Drs Stephen Stansfeld and Bridget Candy at the University of London found that a certain kind of work environment can actually cause clinical depression.[1]

Stansfeld and Candy's study explored several potential factors they suspected could lead to depression in the workplace, including how well teammates worked together, the level of social support, and job security. While these factors are often the topics of water cooler or coffee-break conversation, each proved to have little correlation with mental health.

They did, however, find two particular conditions that predicted a higher likelihood of developing depression at work. 'It doesn't so much matter what you do, but rather the work environment you do it in,'[2] Stansfeld told me.

The first condition involved what the researchers called high 'job strain'. This factor was found in environments where employees were expected to meet high expectations, yet lacked the ability to control the outcomes. Stansfeld added that this strain can be felt in white-collar as well as blue-collar jobs, and likened the feeling to working on a factory production line without a way to adjust the pace of production, even when things go wrong. Like Lucille Ball working in the chocolate factory in the classic episode of *I Love Lucy*, office workers can experience job strain from emails or assignments rushing by like unwrapped chocolates zooming along a conveyor belt.

The second factor that correlates with workplace depression is an environment with an 'effort–reward imbalance', in which workers don't see much return for their hard work, be it through increased pay or recognition. At the heart of both job strain and effort–reward imbalance, according to Stansfeld, is a lack of control.

Depression costs the US economy over $51 billion annually in absenteeism, according to Mental Health America,[3] yet that number doesn't even scratch the surface of the lost potential of millions of Americans who suffer at work without a medical diagnosis. Furthermore, it also doesn't account for the mild depression-like symptoms caused by unhealthy work environments that lead to unwanted consequences, such as distraction. Because we turn to our devices to escape discomfort, we often reach for our tech tools to feel better when we experience a lack of control. Checking email or chiming in on a group chat thread provides the feeling of being productive, regardless of whether our actions are actually making things better.

Technology is not the root cause of distraction at work. The problem goes much deeper.

Dr Leslie Perlow, a consultant turned professor at Harvard Business School, led an extensive four-year study documented in her book *Sleeping with Your Smartphone*.[4] In her book she writes of managers at the Boston Consulting Group (BCG), a leading strategy consulting firm, who perpetuated the high expectations and low-control work culture associated with mental illness.

For example, Perlow describes a project led by two partners at the firm with opposing work styles. One of them was an early riser, while the other was a night owl. Like parents embroiled in a nasty divorce, the two were rarely in the same room and would communicate through their team. A consultant on the team recalls:

> The more junior partner was continually asking us to expand and add things, so we would end up with forty- to sixty-page slide decks for the weekly meetings. The senior partner would wonder why we were all in the red zone [working more than sixty-five hours per week] and say, 'Why do you have all the stuff?' ... One partner was up late and would send us changes at 11 p.m., the other was up early sending emails at 6 a.m ... We were getting it on both ends.[5]

These anecdotes may be unique, but the problems they highlight are not. Employees doing their duty and trying to please their managers often feel unable to change the way things function. As a consultant Perlow interviewed said, 'Partners like hearing "yes", more than they like hearing "no", and I'm trying to give them what they want.'

If a manager sent an email at an hour traditionally reserved for one's family or sleep, it would be read and replied to. If a manager wanted a meeting to discuss whatever they felt needed discussing, despite other pressing matters, the team would drop everything and attend the meeting. If a manager felt the team needed to work late (irrespective of existing personal plans of the team's members), well, you can guess what happened.

The addition of technology to this corrosive culture made things worse. Perlow described how 'pressure to be [always] on gets amplified' in what she calls the 'cycle of responsiveness'. She writes, 'The pressure to be on usually stems from some seemingly legitimate reason, such as requests from clients or customers or teammates in different time zones.' As a result, employees 'begin adjusting to these demands – adapting the technology they use, altering their daily schedules, the way they work, even the way they live their lives and interact with their families and friends – to be better able to meet the increased demands on their time.'

How workplace technology drives us crazy

1. 'People here are always connected.'

4. Increasing expectations to be always on

2. Reducing control over one's time

3. 'To get ahead I need to be always available.'

Source: Inspired by Leslie Perlow book, Sleeping With Your Cell Phone

While technology perpetuates a vicious 'cycle of responsiveness', a dysfunctional culture is its fuel.

Increased accessibility comes at a high price. Answering emails during your child's soccer game trains colleagues to expect quick responses during times that were previously off-limits; as a result, requests from the office mutate personal or family time into work time.

More requests mean more pressure to respond, as email inboxes overflow and Slack messages continue to pour in. Soon, a culture of always-on responsiveness becomes the office norm – exactly as it did at BCG.

The cycle of responsiveness is caused by a cascade of consequences. Technology such as the mobile phone and Slack may fuel the cycle, but the technology itself isn't the source of the problem; rather, overuse is a symptom.

Dysfunctional work culture is the real culprit.

Happily, the BCG story has a good ending. Once Perlow realised the source of the problem, she helped the company change its toxic culture. In the process, she revealed that if a company was unable to address an issue like technology overuse, it was probably also concealing all sorts of deeper problems. In the following chapters in this section, I'll expand on what Perlow did to help BCG and what you can do to change the culture of distraction at your workplace.

▌REMEMBER THIS:

- Jobs where employees encounter high expectations and low control have been shown to lead to symptoms of depression.
- **Depression-like symptoms are painful.** When people feel bad, they use distractions to avoid their pain and regain a sense of control.
- Tech overuse at work is a symptom of a dysfunctional company culture.
- More tech use makes the underlying problems worse, perpetuating a 'cycle of responsiveness'.

27

Fixing Distraction Is a Test
of Company Culture

When Dr Leslie Perlow began her research at the Boston Consulting Group, she was well aware of the firm's round-the-clock reputation. Her interviews with BCG staff soon revealed why the company struggled with an employee retention problem.* Lack of control over their schedules and the expectation that they would be constantly connected were major reasons why people left the firm.

To tackle the issue, Perlow came up with a simple proposition: if everyone who worked at BCG hated the always-on lifestyle, why not try to give consultants at least a 'single predictable night off a week'? This would give people time away from phone calls and email notifications and allow them to make plans without the fear of being pulled back into work.[1]

Perlow ran the idea by George Martin, the managing partner of the Boston office, who promptly told her to keep her hands off *his* teams. However, perhaps in an attempt to get the curious researcher out of his hair, he gave her permission to 'wander

*My first job out of college was at BCG, well before Perlow's work at the company. I did not stay at the firm for long.

around the office' and look for 'another partner who might be willing'. Perlow finally found a young partner named Doug who had two small children at home and a third on the way. Doug was struggling to balance his own work life and agreed to let his team serve as the guinea pigs in Perlow's experiment. Starting with Doug and the people he managed, Perlow proposed the challenge and began studying how the team went about finding a way to let everyone unplug.

First, Perlow confirmed that one night off per week was a universally desired goal for everyone on the team. After hearing a resounding 'Yes!' the team was left to figure out exactly how they would structure their work days to achieve the goal. The team met regularly to discuss roadblocks that were preventing them from achieving the 'one night off' mission and came up with new practices they'd need to implement to make it happen.

For years, BCG consultants had heard countless reasons why they had to be accessible at all hours. 'We're in the service business', 'We work across time zones' and 'What if a client needs us?' were common responses that cut off attempts to find better ways of working. However, once they had an opportunity to openly discuss the problem, Doug's team discovered there were many simple solutions.

A common workplace dilemma that was often dismissed as 'the way things had to be' could be solved if people had a safe space to talk about the issue, without fear of being labelled as 'lazy' for wanting to turn off their phones and computers for a few hours.

To Perlow's surprise, these meetings yielded far greater benefits than she expected, addressing topics well beyond disconnecting from technology. The meetings to discuss predictable time off 'made it okay for people to speak openly', which, in Perlow's words, 'was a big deal.'

What had started as a discussion about disconnecting became a forum for open dialogue.

Team members found themselves questioning other company norms. Having a place to ask, 'Why do things have to be this way?' gave them a forum to generate new ideas. 'There was no taboo,' one consultant said. 'You could talk about anything.' The senior members of the team 'did not always agree, but it was okay to bring anything up'.

Managers also found a venue to explain their larger objectives and strategy – topics that had previously been brushed aside when things got busy. With a clearer view of how their work contributed to a larger vision, team members felt more empowered and able to affect the outcome of their projects. As ideas flowed, meetings became natural opportunities to praise team members for their contributions, raise concerns and voice issues that previously could not be addressed elsewhere.

Embracing Perlow's challenge stopped the cycle of responsiveness. Rather than blaming the technology for their problems, the teams reflected on the reasons behind its overuse. The toxic, always-on culture was no longer accepted as the way things had to be but was seen as another challenge that could be overcome once people were allowed to address it openly.

What began as a challenge to find a way to let members of one team disconnect one night per week profoundly changed the working culture at BCG. Once the epitome of the sort of workplace environment associated with higher rates of depression, as identified in Stansfeld and Candy's study, BCG began a company-wide transformation.

Today, teams throughout the firm (including George Martin's Boston office) have adopted the practice of conducting regular meetings to ensure everyone has time to disconnect. More importantly, providing a safe place for open dialogue about all sorts of issues increased employees' sense of control and turned out to be an unexpected way of improving job satisfaction and staff retention. When team members were given what they needed to flourish, they found ways to address the real problems that had been holding them, and their company, back.

Companies consistently confuse the disease of bad culture with symptoms like tech overuse and high employee turnover.

The problem Perlow discovered at BCG plagues organisations of every size and in every industry. Google recently set out to understand the drivers of employee retention and the quality of team outcomes. The search giant announced the results of a two-year study to understand, once and for all, the answer to the question 'What makes a Google team effective?'[2]

Heading into the study, the research team was fairly confident of what they would find: that teams are most effective when they are composed of great people. As Julia Rozovsky, a researcher on the project, wrote:

Take one Rhodes Scholar, two extroverts, one engineer who rocks at AngularJS, and a Ph.D. *Voilà.* Dream team assembled, right? We were dead wrong. Who is on a team matters less than how the team members interact, structure their work and view their contributions.

The researchers found 'five key dynamics that set successful teams apart', including dependability, structure and clarity, meaning, and impact. However, 'far and away the most important of the five dynamics we found ... the underpinning of the other four' was something called 'psychological safety'. Rozovsky continued:

> Individuals on teams with higher psychological safety are less likely to leave Google, they're more likely to harness the power of diverse ideas from their teammates, they bring in more revenue, and they're rated as effective twice as often by executives.

The term 'psychological safety' was coined by Dr Amy Edmondson, an organisational behavioural scientist at Harvard. Edmondson defines psychological safety as 'a belief that one will not be punished or humiliated for speaking up with ideas, questions, concerns, or mistakes'.[3] Speaking up sounds easy, but if you don't feel psychological safety you'll keep your concerns and ideas to yourself. Rozovsky continues:

> Turns out, we're all reluctant to engage in behaviours that could negatively influence how others perceive our competence, awareness, and positivity. Although this kind of self-protection is a natural strategy in the workplace, it is detrimental to effective teamwork. On the flip side, the safer team members feel with one another, the more likely they are to admit mistakes, to partner, and to take on new roles.

Psychological safety is the antidote to the depression-inducing work environments Stansfeld and Candy found in their study. It's also the magic ingredient the teams at BCG found when they began regular meetings to address the challenge of giving employees predictable time off.

Knowing that your voice matters, and that you're not stuck in an uncaring, unchangeable machine, positively impacts wellbeing.

How does a team, or a company for that matter, create psychological safety? Edmondson provides a three-step answer in her TEDx talk:[4]

Step 1: 'Frame the work as a learning problem, not an execution problem.' Because the future is uncertain, emphasise that 'we've got to have everyone's brains and voices in the game,' says Edmonson.

Step 2: 'Acknowledge your own fallibility.' Managers need to let people know that nobody has all the answers – that we're in this together.

Step 3: Finally, she suggests that leaders must 'Model curiosity and ask lots of questions'.

Edmondson insists that organisations – particularly those operating in conditions of high uncertainty and interdependence among team members – need also to have high levels of motivation and psychological safety, a state she calls the 'learning zone'.

It's in the learning zone that teams perform at their best and it's where they can air concerns without fear of being attacked or fired. It's where they can solve problems, like that of tech overuse and distraction, without being judged as unwilling to carry their share. It's where they can enjoy a company culture that frees them from the nagging internal triggers brought on when they feel a lack of control.

Only when companies give employees a psychologically safe place to air concerns and solve problems together can they solve some of their biggest workplace challenges. Creating an environment where employees can do their best without distraction puts the quality of the organisation's culture to the test. In the next chapter we'll learn from companies that pass with flying colours.

▌ REMEMBER THIS:

- **Don't suffer in silence.** A workplace where people can't talk about technology overuse is also one where people keep other important issues (and insights) to themselves.
- **Knowing that your voice matters is essential.** Teams that foster psychological safety and facilitate regular open discussions about concerns not only have fewer problems with distraction, but also have happier employees and customers.

28

The Indistractable Workplace

If there's one technology that embodies the unreasonable demands of the always-on work culture that pervades so many companies today, it's Slack. The group chat app can make users feel tethered to their devices, often at the expense of doing more important tasks.

More than ten million people log on to Slack every day.[1] The platform's employees, of course, use Slack – they use it a lot. And if distraction is caused by technology, then they should surely suffer the consequences. Surprisingly, according to media reports and Slack employees I spoke with, the company doesn't have that problem.

If you were to walk around Slack's company headquarters in San Francisco, you'd notice a peculiar slogan on the hallway walls. White letters on a bright pink background blare out 'Work hard and go home'. It's not the kind of motto you'd expect to see at a Silicon Valley company that, ironically, makes the very tool many people say keeps them at work, even after they've gone home.

However, at Slack, people know when to log off. According to a 2015 article in *Inc.* magazine that named Slack its Company of the Year, the slogan is more than just talk. By 6.30 p.m., 'Slack's offices have pretty much cleared out'.[2] And according to the article, 'That's how [Slack CEO Stewart] Butterfield wants it.'

Surely Slack employees log back in when they get home, right? Wrong. In fact, they are discouraged from using Slack after they've left work. According to Amir Shevat, Slack's former director of developer relations, people there understand the norm is to know when to disconnect. 'It's not polite to send direct messages after hours or during weekends,' he adds.

Slack's corporate culture is an example of a work environment that hasn't succumbed to the maddening cycle of responsiveness endemic to so many organisations today.

To facilitate focus, Slack's culture goes even deeper than its slogans. Slack management leads by example to encourage employees to take time to disconnect. In an interview with *OpenView*,[3] Bill Macaitis, who served as Slack's chief revenue officer and chief marketing officer, states, 'I'm a big productivity guy … You need to have uninterrupted work time … This is why – whether I'm dealing with Slack or email – I always block off time to go in and check messages and then return to uninterrupted work.' The fact that someone as senior as Macaitis makes uninterrupted work a priority, and goes as far as scheduling time for email and Slack, sends a profound message that exemplifies the principle of 'making time for traction' we covered in Part 2.

Shevat echoes Macaitis's sentiment. At Slack, he said, 'It's okay to be offline.' He is religious about giving his co-workers his complete attention when meeting them in person: 'When I give someone my time, I'm focused 100 per cent and never open a phone during a meeting. That is super-important for me.' By taking steps to remove the buzzes and rings typical of modern meetings, he practises the idea of 'hacking back external triggers' we discussed in Part 3.

Shevat also revealed how Slack employees use a precommitment pact, the kind we discussed in Part 4, to help them stay offline outside office hours. Slack has a 'Do Not Disturb' feature built into the service that users can turn on whenever they want to concentrate on what they really want to do, like doing focused work or being with family or friends. Shevat told me that if an employee tries to send a message when they shouldn't, 'you will get hit by the "Do Not Disturb" feature. If it's after hours, it turns on automatically so you don't get direct messages until you get back to work.'

Most importantly, the culture at Slack ensures employees have a place to discuss their concerns. As Leslie Perlow discovered at BCG, regular meetings were critical for airing employee concerns. Companies that make time to discuss their issues are more likely to foster psychological safety and hear the looming problems employees would otherwise keep to themselves.

As we learned in Part 1, dealing with distraction starts by our understanding what's going on inside us. If internal triggers are crying out for relief, employees will find ways to address them one way or another – healthily or not. Ensuring employees have a forum to voice problems to company leadership helps Slack team members relieve the psychological strain Stansfeld and Candy found in toxic work environments.

But how does a company as big as Slack make sure everyone has a place to feel heard? This is where the company's own technology comes in handy. The group chat tool facilitates the regular discussions needed to foster psychological safety while coming to consensus quickly. How do they do it? While it may seem inconceivable, Shevat credits emojis.

At Slack, he says, there's a channel for everything. 'We have a channel for people who want to get lunch together, a channel for sharing pet photos, even a *Star Wars* channel.' These separate channels not only save others from the sort of off-topic conversations that clog up email and make in-person meetings unbearable – they also give people a safe place to send feedback.

Among the many Slack channels, the ones company leadership takes most seriously are the feedback channels. They are not just for sharing opinions on the latest product release; they are also for sharing thoughts about how to improve as a company. There is a dedicated channel called '#slack-culture' and another where executives invite employees to 'ask me anything' called '#exec-ama'. Shevat says, 'People will post all sorts of suggestions and are encouraged to do so.' There's even a special channel for airing your 'beef' with the company's own product, called '#beef-tweets'. 'Sometimes comments can get very prickly,' Shevat says, but the important thing is that they're aired and heard.

Here's where emojis can come to the rescue. 'Management lets people know they've read their feedback with an eyes emoji. Other times, if something is handled or fixed, someone will respond with a check mark,' Shevat explains. Slack has found a way to let its employees know they're being heard and action is being taken.

Of course, not every conversation at every company should take place in a group chat. Slack also conducts regular All Hands meetings where employees can ask senior management questions directly. No matter the format, giving employees a way to send feedback, and also know it's been heard by someone who can help, lets them know they have a voice. Whether employees' feedback is heard during small-group meetings, like those facilitated by Perlow at BCG, or over group chat channels as at Slack, isn't the point; what matters is that there is an outlet that management cares about, uses and responds to. It is critical to the wellbeing of a company and its employees.

***

There's always a risk when pointing to specific companies as exemplars. Jim Collins' bestsellers, *Good to Great* and *Built to Last*, included profiles of some companies that ended up not lasting very long and others that turned out to be not so great.[4] Certainly, working at Slack and BCG isn't perfect. Some employees I spoke with told me they'd had bad experiences with heavy-handed managers. As one former employee said of Slack, 'They really did try to be a psychologically safe company. It's just that not everyone was equally skilled at maneuvering some of those nuances.' Creating the kind of company where people feel comfortable raising concerns without the fear of getting fired takes work and vigilance.

For now, the strategies of BCG and Slack seem to be successful. Both organisations are loved by their employees and customers; on Glassdoor.com, BCG has been named among the ten 'Best Places to Work' for eight of the past nine years,[5] while Slack has an average anonymous review of 4.8 out of 5 stars, with 95 per cent of employees saying they'd recommend the company to a friend, and 99 per cent approval of the CEO.[6]

It is worth noting that, regardless of future profit margins or returns to shareholders, these companies, at the time of writing, show concern and commitment to helping their employees thrive by giving them the freedom to be indistractable.

★ REMEMBER THIS:

- Indistractable organisations, like Slack and BCG, foster psychological safety, provide a place for open discussions about concerns, and, most importantly, have leaders who exemplify the importance of doing focused work.

Part 6

How to Raise Indistractable Children
(And Why We All Need
Psychological Nutrients)

29

Avoid Convenient Excuses

Society's fear of what a potential distraction like the smartphone is doing to our kids has reached fever pitch. Articles with headlines like 'Have Smartphones Destroyed a Generation?'[1] and 'The Risk Of Teen Depression And Suicide Is Linked To Smartphone Use, Study Says'[2] have, ironically enough, gone viral online.

Dr Jean Twenge, the author of the former article writes, 'It's not an exaggeration to describe iGen as being on the brink of the worst mental-health crisis in decades. Much of this deterioration can be traced to their phones.'[3]

Convinced by the ominous headlines and fed up with their kids' tech distractions, some parents have resorted to extreme measures. A search on YouTube reveals thousands of videos of parents storming into their kids' rooms, unplugging the computers or gaming consoles and smashing the devices to bits in order to teach their kids a lesson.[4] At least, that's their hope.

I can certainly understand parents' feelings of frustration. One of the first things my daughter ever said was, 'iPad time. iPad time!' If we didn't comply quickly, she'd increase the volume until we did, raising our blood pressure and testing our patience. As the years passed, my daughter's relationship with screens evolved, and not always in a good way. She was drawn to spending too much time playing frivolous apps and watching videos.

Now that she's older there are new problems associated with raising a kid in the digital age. On more than one occasion, we've met up with friends and their kids for dinner, only to find ourselves sitting through awkward meals as the kids spent the entire time tap-tap-tapping away at their phones instead of engaging with one another.

As tempting as it may be, destroying a kid's digital device isn't helpful. Surrounded by alarming headlines and negative anecdotes, it's easy to understand why many parents think tech is the source of the trouble with kids these days. But is it? As we've seen in the case in the workplace and in our own lives, there are once again hidden root causes to kids' distraction.

***

My wife and I needed to help our daughter develop a healthy relationship with tech and other potential distractions, but first had to work out what was causing her behaviour. As we've learned throughout this book, simple answers to complex questions are often wrong, and it is far too easy to blame behaviour parents don't like on something other than ourselves.

For example, every parent *obviously* knows children become hyperactive when they eat too much sugar. We've all heard parents blame their kid's brattish behaviour at the birthday party on the ominous 'sugar high'. I must admit I've used that excuse on more than one occasion myself. That is, until I learned that the concept of a 'sugar high' is total scientific bunk. An exhaustive meta-analysis of sixteen studies 'found that sugar does not affect the behaviour or cognitive performance of children'.[5]

Interestingly, though the so-called 'sugar high' is a myth for kids, it does have a real effect on parents. A study found that mothers, when told that their sons were given sugar, rated their child's behaviour as more hyperactive – despite that child having been given a placebo. In fact, videotapes of the mothers' interactions with their sons revealed that they were more likely to trail their children and criticise them when they believed they were 'high' on sugar – again, despite the fact that their sons hadn't eaten any.

Another classic excuse in the parental toolkit of blame deflection is the 'common knowledge' that teens are rebellious by nature. Everyone knows that teenagers act horribly towards their parents because their raging hormones and underdeveloped brains *make* them act that way. Wrong.

Studies have found that teenagers in many societies, particularly pre-industrialised ones, don't act especially rebelliously and, conversely, spend 'almost all their time with adults'.[6] In an article, 'The Myth of the Teen Brain', Robert Epstein writes, 'many historians note that through most of recorded human history, the teen years were a relatively peaceful time of transition to adulthood'.[7] Apparently, our teenagers' brains are fine – it is *our* brains that are underdeveloped.

Innovations and new technologies are another frequent target of blame. In 1474 Venetian monk and scribe Filippo di Strata issued a polemic against another handheld information device, stating, 'the printing press a whore'. An 1883 medical journal attributed rising rates of suicide and homicide to the new 'educational craze', proclaiming 'insanity is increasing

. . . with education' and that education would 'exhaust the children's brains and nervous systems'.[8] In 1936, kids were said to have 'developed the habit of dividing attention between the humdrum preparation of their school assignments and the compelling excitement of the [radio] loudspeaker,' according to the music magazine *Gramophone*.[9]

It seems hard to believe that these benign developments scared anyone, but leaps in technological innovation are often followed by moral panic. 'Each successive historical age has ardently believed that an unprecedented "crisis" in youth behaviour is taking place,' the Oxford historian Dr Abigail Wills wrote for the BBC. 'We are not unique; our fears do not differ significantly from those of our predecessors.'[10]

When it comes to the undesirable behaviour of children today, convenient myths about devices are just as dubious as the blame parents deflect on to sugar highs, underdeveloped teen brains and other technologies like the book and the radio.

Many experts believe the discussion regarding whether tech is harmful is more nuanced than alarmists let on.

In a rebuttal to the article that claimed children are on the brink of the worst mental health crisis in decades, Dr Sarah Rose Cavanagh wrote in *Psychology Today*, 'the data the author chooses to present are cherry-picked, by which I mean she reviews only those studies that support her idea and ignores studies that suggest that screen use is NOT associated with outcomes like depression and loneliness'.[11]

One of many studies not cherry-picked was conducted by Dr Christopher Ferguson and published in *Psychiatric Quarterly*. It found only a negligible relationship between screen time and depression; Ferguson wrote in an article in *Science Daily*, 'Although an "everything in moderation" message when discussing screen time with parents may be most productive, our results do not support a strong focus on screen time as a preventative measure for youth problem behaviours.'[12] As is so often the case, the devil is in the digital details.

A closer read of the studies linking screen time with depression finds correlation only with extreme amounts of time spent online. Teenage girls who spent over five hours per day online tended to have more depressive or suicidal thoughts, but common sense would have us ask whether the kids who have a propensity to spend excessive amounts of time online might also have other problems in their lives. Perhaps five hours a day on any form of media is a symptom of a larger problem.

In fact, the same study found that kids who spent two hours or less online per day did not have higher rates of depression and anxiety compared to controls. A study conducted by Dr Andrew Przybylski at the Oxford Internet Institute found that mental wellbeing actually *increased* with moderate amounts of screen time.[13] 'Even at exceptional levels, we're talking about a very small impact,' stated Przybylski.[14] 'It's about a third as bad as missing breakfast or not getting eight hours' sleep.'

***

When kids act in ways we don't like, parents have always craved an answer to the question 'why is my kid acting this way?' There's certainty in a scapegoat and we often cling to simple answers because they serve a story we want to believe – that kids do strange things because of something outside our control, which means that those behaviours are not really their (or our) fault.

Of course, technology plays a role. Smartphone apps and video games are designed to be engaging, just as sugar is meant to be delicious. But like the parent who blames a 'sugar high' for their kid's bad behaviour, blaming devices is a superficial answer to a deep question. Easy answers mean we don't have to look into the dark and complex truth underlying why kids behave the way they do. But we can't fix the problem unless we look at it clearly, free of media-hyped myths, to understand the root causes.

Parents don't need to believe tech is evil to help kids manage distraction.

Learning to become indistractable is a skill that will serve our children no matter what life path they pursue or what forms distraction takes. If we are going to help our kids take responsibility for their choices, we need to stop making convenient excuses for them – and for ourselves. In the following chapters in Part 6, we're going to understand the deeper psychology driving some kids to overuse their devices and learn smart ways to help them overcome distraction.

★ REMEMBER THIS:

- **Stop deflecting blame.** When kids don't act the way parents want them to, it's natural to look for answers that help parents divert responsibility.
- **Techno-panics are nothing new.** From the book, to the radio, to video games, the history of parenting is strewn with examples of moral panic over things that supposedly make kids act in strange ways.
- **Tech isn't evil.** Used in the right way and in the right amounts, kids' tech use can be beneficial, while too much (or too little) can have slightly harmful effects.
- **Teach kids to be indistractable.** Teaching children how to manage distraction will benefit them throughout their lives.

30

Understand Their Internal Triggers

Dr Richard Ryan and his colleague Dr Edward Deci are two of the most frequently cited researchers in the world on the drivers of human behaviour. Their 'self-determination theory' is widely regarded as the backbone of psychological wellbeing, and countless studies have supported their conclusions since they began research in the 1970s.[1]

Just as the human body requires three macronutrients (protein, carbohydrates and fat) to run properly, Ryan and Deci proposed that the human psyche needs three things to flourish: autonomy, competence and relatedness. When the body is starved, it elicits hunger pangs; when the psyche is undernourished, it produces anxiety, restlessness and other symptoms that something is missing.

When kids aren't getting the psychological nutrients they need, self-determination theory explains why they might overdo unhealthy behaviours, such as spending too much time in front of screens. Ryan believes the cause has less to do with the devices and more to do with why certain kids are susceptible to distraction in the first place.

Without sufficient amounts of autonomy, competence and relatedness, kids turn to distractions for psychological nourishment.

FIRST, KIDS NEED AUTONOMY – VOLITION AND FREEDOM OF CONTROL OVER THEIR CHOICES.

Maricela Correa-Chávez and Barbara Rogoff, professors at the University of California Santa Cruz, conducted an experiment in which two children were brought into a room where an adult taught one of them how to build a toy while the other one waited.[2] The study was designed to observe what the non-participating child, the observer, would do while they waited. In America, most of the observer children did what you'd expect them to do: they shuffled in their seats, stared at the floor and generally showed signs of disinterest. One impatient boy even pretended a toy was a bomb and threw his hands in the air to mimic an explosion, making loud disruptive noises to match the carnage. In contrast, the researchers found that Mayan children from Guatemala concentrated on what the other child was learning and sat still in their chairs as the adult taught the other child.

Overall, the study found that American children could focus for only half as long as Mayan kids. Even more interesting was the finding that the Mayan children with less exposure to formal education 'showed more sustained attention and learning than their counterparts from Mayan families with extensive involvement in Western schooling'. In other words, less schooling meant more focus. How could that be?

Psychologist Dr Suzanne Gaskins has studied Mayan villages for decades and told *NPR* that Mayan parents give their kids a tremendous amount of freedom. 'Rather than having the mom set the goal – and then having to offer enticements and rewards to reach that goal – the child is setting the goal. Then the parents support that goal however they can,' Gaskins said. Mayan parents 'feel very strongly that every child knows best what they want and that goals can be achieved only when a child wants it.'[3]

Most formal schooling in America and similar industrialised countries, on the other hand, is the antithesis of a place where kids have the autonomy to make their own choices. According to Rogoff, 'It may be the case that [some American] children give up control of their attention when it's always managed by an adult.'[4] In other words, kids can become conditioned to *lose control* of their attention and become highly distractable as a result.

Ryan's research reveals exactly where we lose kids' attention. 'Whenever children enter middle school, whenever they start leaving home-based classrooms and go into the more police-state style of schools, where bells are ringing, detentions are happening, punishment is occurring, they're learning right then that this is not an intrinsically motivating environment,' he says.[5] Robert Epstein, the researcher who wrote 'The Myth of the Teen Brain' in *Scientific American*, draws a similar conclusion. 'Surveys I have conducted show that teens in the US are subjected to more than ten times as many restrictions as are mainstream adults, twice as many restrictions as active-duty US Marines, and even twice as many restrictions as incarcerated felons,' he wrote.[6]

While such a restrictive environment isn't every American student's experience, it's clear why so many struggle to stay motivated in the classroom: their need for autonomy to explore their interests is unfulfilled. 'We're doing a lot of controlling them in their school environments and it's no surprise that they should then want to turn to an environment where they can feel a lot of agency and a lot of autonomy in what they're doing,' Ryan says. 'We think of [tech use] as kind of an evil in the world, but it's an evil we have created a gravitational pull around by the alternatives we've set up.'[7]

Unlike their offline lives, kids have a tremendous amount of freedom online; they have the autonomy to call the shots and experiment with creative strategies to solve problems. 'In internet spaces, there tend to be myriad choices and opportunities, and a lot less adult control and surveillance,' says Ryan. 'One can thus feel freedom, competence and connection online, especially when the teenager's contrasting environments are overly controlling, restrictive or understimulating.'

Ironically, when parents grow concerned about how much time their kids spend online, they often impose even more rules – a response that tends to backfire. Instead of more ways to limit your kids' autonomy, Ryan advises seeking to understand the underlying needs and associated internal triggers driving them to digital distraction. 'What we've found is that parents who address internet use or screen time with kids in an autonomy-supported way have kids who are more self-regulated with respect to it, so less likely to use screen time for excessive hours,' he says.

SECOND, CHILDREN STRIVE FOR COMPETENCE —
MASTERY, PROGRESSION, ACHIEVEMENT AND GROWTH.

Think about something you're good at: your ability to present on stage, pull together a delicious meal, or parallel park in the tightest of spaces. Competence feels good, and that feeling grows alongside your ability.

Unfortunately, the joy of progress in the classroom is on the wane among kids today. Ryan warns, 'We're giving messages of "you're not competent at what you're doing at school", to so many kids.' He points to the rise of standardised testing as part of the problem. 'It's destroying classroom teaching practices, it's destroying the self-esteem of so many kids and it's killing their learning and motivation.'

'Kids are so different, and their developmental rates are so variable,' Ryan continues. However, by design, standardised tests don't account for those differences. If a child isn't doing well in school and doesn't get the necessary individualised support, they start to believe that achieving competence is impossible, so they stop trying. In the absence of competency in the classroom, kids turn to other outlets in order to experience the feeling of growth and development. Companies making games, apps and other potential distractions are happy to fill that void by selling ready-made solutions for the 'psychological nutrients' kids lack.

Tech makers know how much consumers enjoy levelling up, gaining more followers or getting likes – those accomplishments provide the fast feedback of achievement that feels good. According to Ryan, when children spend their time in school doing something they don't enjoy, don't value and don't see potential for improvement in, 'it should be no surprise to us that at night-time [they] would rather turn to an activity where they can feel a lot of competence'.

THIRD, THEY SEEK RELATEDNESS – FEELING IMPORTANT TO OTHERS AND THAT OTHERS ARE IMPORTANT TO US.
Spending time with peers has always been a formative part of growing up. For kids, much of the opportunity to develop social skills centres around chances to play with others. In today's world, however, teens increasingly experience social interactions in virtual environments because doing so in the real world is inconvenient or off-limits.

The very nature of play is rapidly changing. Remember playing pick-up games at the basketball court, hanging out at the mall on weekends or simply roaming around the neighbourhood until you found a friend? Sadly, spontaneous socialising simply isn't happening as much as it used to.

As Dr Peter Gray, who has studied the decline of play in America, told *The Atlantic*, 'It is hard to find groups of children outdoors at all, and, if you do find them, they are likely to be wearing uniforms and following the directions of coaches.'[8]

Whereas previous generations were simply allowed to play after school and form close social bonds, many children today are raised by parents who restrict outdoor play, due to 'child predators, road traffic, and bullies', according to a survey of parents in the same *Atlantic* article.[9] These concerns were mentioned despite the fact that kids today are statistically the safest generation in American history.[10] Unfortunately, irrational fears lead to a downward spiral that leaves many kids with no choice but to stay indoors, attend structured programmes or rely on technology to find and connect with others.

In many ways, connections in digital environments can be very positive. A child who is bullied at school can reach out for help from supportive online friends; a teenager struggling with their sexuality can find support from someone on the other side of the country; and a kid who feels shy at school can be a hero among their gaming friends from all corners of the world. 'What the data show,' says Ryan, 'is that kids who aren't feeling relatedness, who are feeling isolated or excluded in school are going to be more drawn to media where they can get connections with other people and find subgroups they can identify with. So that's both a plus and a minus.'[11]

The loss of in-person play has real costs according to Gray, given that 'learning to get along and cooperate with others as equals may be the most crucial evolutionary function of human social play'. He sees it as 'both a consequence and a cause of the increased social isolation and loneliness in the culture'. Long before studies correlated screen time with rising rates of depression, Gray identified a much bigger trend that goes back over sixty years:

> Since about 1955 ... children's free play has been continually declining, at least partly because adults have exerted ever-increasing control over children's activities. [A]s a society, we have come to the conclusion that to protect children from danger and to educate them, we must deprive them of the very activity that makes them happiest and place them for ever more hours in settings where they

are more or less continually directed and evaluated by adults, setting[s] almost designed to produce anxiety and depression.[12]

***

When considering the state of modern childhood, Ryan believes many kids aren't getting enough of the three essential psychological nutrients – autonomy, competence and relatedness – in their offline lives. Not surprisingly, our kids go looking for substitutions online. 'We call this the "need density hypothesis"', says Ryan.[13] 'The more you're not getting needs satisfied in life, reciprocally, the more you're going to get them satisfied in virtual realities.'[14]

Ryan's research leads him to believe that 'overuse [of technology] is a symptom, one indicative of some emptiness in other areas of life, like school and home'. When these three needs are met, people are more motivated, perform better, persist longer and exhibit greater creativity.

Ryan isn't against setting limits on tech use, but thinks such limits should be set with the child, and not arbitrarily enforced because you think you know best. 'Part of what you want your kid to get from that is not just less screen time, but an understanding of why,' he says. The more you talk with your kids about the costs of too much tech use and the more you make decisions *with* them, as opposed to *for* them, the more willing they will be to listen to your guidance.

We can start by sharing some of the coping and reimagining tactics we learned in Part 1. Let your children know what you're doing differently in your own life to manage distraction; being vulnerable and showing kids that we understand their struggle and face similar challenges helps build trust. Just as we saw in the previous section how good bosses model disconnecting from distraction, parents should model how to be indistractable.

We may also want to consider providing real-world opportunities for children to find the autonomy, competence and relatedness they need. Easing up on structured academic or athletic activities and giving them more time for free play may help them find the connections they otherwise look for online.

We can't solve all our kids' troubles, and nor should we attempt to, but we *can* try to understand their struggles better through the lens of their psychological needs. Knowing what's really driving their overuse of technology is the first step to helping kids build resilience instead of escaping discomfort through distraction. Once our kids feel understood, they can begin planning how best to spend their time.

REMEMBER THIS:

- **Internal triggers drive behaviour.** To understand how to help kids manage distraction, we need to start by understanding the source of the problem.
- **Our kids need psychological nutrients.** According to a widely accepted theory of human motivation, all people need three things to thrive: a sense of autonomy, competence and relatedness.
- **Distractions satisfy deficiencies.** When our kids' psychological needs are not met in the real world, they go looking for satisfaction – often in virtual environments.
- **Kids need alternatives.** Parents and guardians can take steps to help kids find a balance between their online and offline worlds by providing more offline opportunities to find autonomy, competence and relatedness.
- **The four-part Indistractable Model is valuable for kids as well.** Teach them methods for handling distraction and, most importantly, model being indistractable yourself.

31

Make Time for Traction Together

When it comes to helping our kids manage distraction, it's important to make the conversation about people rather than tech. That's according to Lori Getz, the founder of Cyber Education Consultants, which hosts internet safety workshops in schools. It's a lesson she learned in her own childhood.

Getz got her first phone (a corded one for her room) as a teenager. The moment she got it, she closed the door and spent the entire weekend locked in her room, talking with friends instead of spending time with her family. When she got home from school the next Monday, her door had been taken off its hinges. 'It's not the phone's fault you're behaving like an a-hole,' her father chided her. 'You closed the door and you closed all of us out.'

While Getz doesn't recommend her father's aggressive tactics or tone, his focus on the effect her behaviour had on others rather than the phone itself proved instructive. 'Make [the conversation] about how you're treating and interacting with the people around you', she advises, as opposed to blaming the tool.[1]

When it comes to how we spend time together as a family, the important thing is to define what constitutes traction versus distraction. A recent Getz family vacation put her theory to the test. Her six- and eleven-year-old daughters asked if they could use their phones during the two-hour ride from Sacramento to Truckee. Motivated by a desire to ease the monotony of the ride as well as the opportunity for a quiet conversation with her husband, Getz agreed. The device time made the long drive easier, but later in the trip Getz noticed her daughters started turning to their devices a bit too much.

The girls' tech overuse came to a head when Getz returned from a run to find her kids glued to their screens. Neither was ready to leave for their family outing, as had been agreed upon. Rather than losing her cool and punitively announcing strict house rules around the kids' use of devices, Getz decided it was time for a family talk.

During the family huddle, they all confirmed their desire to spend quality time together (aka traction). By agreeing upon how they wanted to spend their time and what needed to get done, it became clear that doing anything else was a distraction interfering with their plans. They decided as a family that they could use their devices only *after* they were 100 per cent ready to go.

Getz acknowledges that admitting you don't have all the answers is a great way to involve the kids in finding new solutions. 'We're all figuring it out as we go along,' she says. Getz wants her daughters to continue to ask themselves questions to self-monitor and self-regulate their behaviour: 'Is my behaviour working for me? Am I proud of myself, in the way I'm behaving?' she asks them to ask themselves. 'I work with a lot of teenagers who will often tell me that they don't want to be distracted, they don't want to be sucked into all this stuff, but they just don't know how to stop.'

To help children learn self-regulation, we must teach them how to make time for traction. We can encourage regular discussions about our values and theirs and teach them how to set aside time to be the people they want to be. Keep in mind that, while it's easy for us to think 'kids have all the time in the world', it's important to remember they have their own priorities within each of *their* life domains.

Working with our kids to create a values-based schedule can help them make time for their personal health and wellness domain, ensuring ample time for rest, hygiene, exercise and proper nourishment. For example, while my wife and I don't enforce a strict bedtime for our daughter, we made it a point to expose her to research findings showing the importance of ample sleep during adolescence. Once *she* realised that sleep was important to her wellbeing, it didn't take much for her to conclude that screen time after 9 p.m. on a school night was a bad idea – a distraction from her value of staying healthy. As you've guessed, she timeboxed periods of rest in her day. While she may occasionally find herself deviating from this evening appointment with her pillow, having it in her schedule provides her with a self-imposed guideline to self-monitor, self-regulate and, ultimately, live out her values.

When it comes to the 'work' domain in kids' lives, for the typical child work is synonymous with school-related responsibilities and household chores. While school schedules provide a timetable for a child's daytime hours, how they spend their time after school can be a source of disagreement and frustration.

Without a clear plan, many kids are left to make impulsive decisions that often involve digital distraction.

I recently had coffee with a friend who is the mother of twin teenage boys. She bemoaned the mind-altering influence of her kids' obsession with the latest techno-villain: the online game *Fortnite*. 'They can't stop!' she told me. She was convinced the game was addictive and her kids were junkies. Every evening saw fights to get them to stop playing and finish their homework, she said. Exasperated, she asked me what I thought she should do.

My advice involved a few unorthodox ideas. First, I advised her to have a conversation with her sons and to listen to them without passing judgement. Potential questions to ask included the following: is keeping up with their school work consistent with their values? Do they know why they are being asked to do their homework? What are the consequences of not doing their assignments? Are they okay with those consequences, both short term (getting a bad grade) and long term (settling for a low-skilled job)?

Without their agreement that school work mattered to them, forcing them to do something they didn't want to do amounted to coercion and would only breed resentment. 'But if I don't hound my kids, they'll fail,' she objected. 'So?' I asked. 'If the only reason they study is to get you off their backs, what will they do when they get to college or start a job and you're not around? Maybe they need to know what failure feels like sooner rather than later?' I advised her that teenagers are generally old enough to make decisions about how they spend their time; if that means flunking a test, then so be it. Coercion may be a band-aid solution, but it is certainly not a remedy.

Next, I proposed she ask them to suggest how much time they'd like to spend on various activities such as studying, being with family or friends or playing *Fortnite*. I warned that while she may not like her kids' answers, it's important to honour their input. The goal here is to teach them to spend their time mindfully by reserving a place for important activities on their weekly schedules. Remember, their schedules (like ours) should be assessed and adjusted weekly to ensure that their time is spent living out their values.

Playing *Fortnite*, for instance, is fine if the time has been allocated to it in advance. With a timeboxed schedule that includes time for digital devices, kids know that they'll have time to do the things they enjoy. I advised her to change the context of their family conversations around tech – from her screaming 'No!' to teaching her kids to tell themselves, 'Not yet'.

Empowering children with the autonomy to control their own time is a tremendous gift. Even if they fail from time to time, failure is part of the learning process.

Finally, I advised her to make sure her kids' days include plenty of time for play, both with their friends and with their parents. Her boys were using *Fortnite* to have fun with their buddies and would continue to play online without an offline alternative. If we want our kids to fulfil their need for relatedness offline, they need time to build face-to-face friendships outside school. These relationships should be free from the pressure of coaches, teachers and parents telling them what to do.

Unfortunately, for the typical child these days, playtime won't happen unless it's scheduled.

Conscientious parents can bring back playtime for kids of all ages by deliberately making time for it in their weekly schedules. Seek out other parents who understand the importance of unstructured play and schedule regular get-togethers to let the kids hang out, just as you would make time for a jog in the park or a jam session in the garage. Research studies overwhelmingly support the importance of unstructured playtime for kids' ability to focus and to develop capacity for social interactions. Given that, unstructured play is arguably their most important extracurricular activity.[2]

In addition to helping kids make time for unstructured play, we also need to carve out time for them to spend time with us, their parents. For example, scheduling family meals is perhaps the single most important thing parents and kids can do together. Studies demonstrate that children who eat regularly with their families show lower rates of drug use, depression, school problems and eating disorders.[3] Unfortunately, many families miss meals together because they 'play it by ear', a strategy that often leaves everyone eating alone on their own schedules. Hence, it's better to set aside an evening, even if only once a week, for a device-free family meal. As our kids develop, we can invite them to shape these family meal experiences by suggesting menu themes like 'Finger-Food Fridays', cooking together or contributing conversation topics.

As a family, play *can* and *should* extend beyond mealtimes. In my household, we've established a weekly 'Sunday Funday', where we rotate the responsibility for planning a three-hour activity. When it's my turn, I might take the family to the park for a long conversation while we walk. When it's her turn to pick, my daughter typically ask to play a board game. My wife often proposes a trip to a local farmers' market to discover and sample new foods. Whatever the choice, the idea is to set aside time together regularly to feed our need for relatedness.

While we must be prepared to make adjustments to our family schedule, we need to involve our kids in setting our routines and honouring our commitments to each other. Teaching them to make their own schedules and being indistractable together helps us pass on our values.

★ REMEMBER THIS:

- **Teach traction.** With so many potential distractions in kids' lives, teaching them how to make time for traction is critical.
- **Just as with our own timeboxed schedules, kids can learn how to make time for what's important to them.** In the absence of making their own plans in advance, kids will turn to distractions.
- **It's okay to let your kids fail.** Failure is how we learn. Show kids how to adjust their schedules to make time to live up to their values.

32

Help Them with External Triggers

After understanding the internal triggers driving kids to distraction and helping them create a schedule using the timeboxing technique, the next step is to examine the external triggers in their lives.

In many ways it's easy to blame the explosion of unwelcome cues tugging at our kids' attention. With their phones buzzing, the television flickering and music blaring into their earbuds, it's difficult to understand how our kids are able to get anything done. Many kids (and adults) pass their days mentally swinging from one thing to the next. Constantly reacting to external triggers, children are left with few opportunities to think deeply and concentrate on anything for long.

According to a Pew Research Center 2015 study in the United States, '95 per cent of teens now report they have a smartphone or access to one'.[1] Not surprisingly, 72 per cent of parents whose kids have a smartphone are concerned they 'pose too much distraction'.[2]

In many ways it is parents and guardians who have enabled this situation. After all, it is *we* who gave permission and often provided the funds to purchase the distracting devices we've come to resent. We've bowed to our kids' demands in ways that may not benefit them or our households.

Many parents don't consider whether their child is ready for a device with potentially damaging consequences and give in to the protestation that 'Everyone in my class has a smartphone and an Instagram account.'

As parents we often forget that a kid wanting something 'really, really badly' is not a good enough reason.

Imagine a young child is standing at the edge of a swimming pool while their friends are all playing in the water and having a great time. The child desperately wants to jump in, but you're not sure they know how to swim. What would you do?

We know swimming pools can be very dangerous, but, despite the risks, we wouldn't keep our children from enjoying the water forever. Rather, once they are old enough, we'd make sure they learned to swim. Even after they'd mastered the basics, we'd keep an eye on them until we were confident about their ability to enjoy the pool safely.

In fact, we can easily think of a host of activities we wouldn't let our kids experience before they're ready: reading certain books, watching age-inappropriate films, driving a car, drinking alcohol and, of course, using digital devices – each comes in its own time, not whenever a kid says so. Exploring the world and navigating its risks is an important part of growing up, but giving a kid a smartphone or other gadgetry before they have the faculties to use it properly is just as irresponsible as letting them jump head first into a pool before they can swim.

Many parents justify handing over smartphones in exchange for the peace of mind of knowing they can contact their children at any time, but, unfortunately, they often find they've given their child too much, too soon. The swimming pool analogy is appropriate here. When a child is learning to enjoy the water, they start in the shallow end. Perhaps they wear armbands or use a kickboard to help them get comfortable with the water. Only later, when they have demonstrated their competence, are they free to swim on their own.

Instead of giving our kids a fully functioning pinging and dinging smartphone, it's better to start with a feature phone that only makes calls and sends text messages. Such a phone can be purchased for less than $25 and does not come with the apps that can distract a child with external triggers.[3] If location tracking is a priority, a GPS-enabled wristwatch like the GizmoWatch keeps track of kids through an app on parents' phones but only allows incoming and outgoing calls to and from select numbers.[4]

As kids get older, a good test of whether they are ready for a particular device is their ability to understand and use the built-in settings for turning off external triggers.

Do they know how to use the 'Do Not Disturb' feature? Do they know how to set their phones to automatically turn off notifications when their schedule demands concentration? Are they able to place their phones out of sight and out of mind during family time or when friends come over? If not, they're not ready, and they need to take a few more 'swimming lessons', so to speak.

Though we parents tend to fixate on the latest technology craze, we often forget about older technologies, which can be just as much of a problem. There's little justification for allowing kids to have a television, laptop or any other potentially distracting external trigger in their rooms; these screens should be kept in communal areas. It is too much to expect our kids to manage the temptation to overuse these devices on their own, particularly in the absence of parental oversight.

Kids also need plenty of sleep, and anything that flickers, beeps or buzzes during the night is a distraction. Anya Kamenetz, the author of *The Art of Screen Time*, writes that making sure kids get enough sleep is 'the one issue with the most incontrovertible evidence'.[5] Kamenetz strongly advises that 'screens and sleep don't mix' and implores parents to keep all digital devices out of kids' rooms at night-time and to shut down screens at least an hour before bedtime.

It's equally important to help our kids remove unwanted external triggers during activities like homework, chores, mealtime, playtime and hobbies that require sustained attention. Just as you may ask your boss for time to focus at work, parents need to respect kids' scheduled time as well. If they are spending time on homework according to their timeboxed schedules, we must of course minimise distraction. But the same rule applies to scheduled time with their friends or playing video games. If they've made their plans in advance and with intent, it's your job to honour that plan and leave them alone.

Remember the critical question: 'Is this external trigger serving me, or am I serving it?' Sometimes, as parents, *we* can be a source of distraction. The dog barking, the doorbell ringing, Dad's subsequent command to answer the door, Mom's question about the baseball team's game schedule or a sibling's invitation to play – all can interfere with the time scheduled for something else. Though these interruptions seem common, any disturbance at the wrong time is a distraction, and we must do our part to help kids use their time as they planned by removing unwanted external triggers.

★ REMEMBER THIS:

- **Teach them to swim before they dive in.** Like swimming in a pool, children should not be allowed to partake in certain risky behaviours before they are ready.
- **Test for tech readiness.** A good measure of a child's readiness is their ability to manage distraction by using the settings on the device to turn off external triggers.
- **Kids need sleep.** There is little justification for having a television or other potential distractions in a kid's room overnight. Make sure nothing gets in the way of getting good rest.
- **Don't be the unwanted external trigger.** Respect their time and don't interrupt them when they have scheduled time to focus on something, be that work or play.

33

Teach Them to Make Their Own Pacts

When my daughter was five years old and already insisting on 'iPad time' with unrelenting protests, my wife and I knew we had to act. After everyone had calmed down, we did our best to respect her needs in the way Dr Ryan recommends: we explained, as simply as we could, that too much screen time comes at the expense of other things.

As a kindergartner, she was learning to tell the time, so we could explain that there was only so much of it for things she enjoyed. Spending too much time with apps and videos meant less time to play with friends in the park, swim at the community pool, or be with Mom and Dad.

We also explained that the apps and videos on the iPad were made by some very smart people and were intentionally designed to keep her hooked and habitually watching. It's important that our kids understand the motives of the gaming companies and social networks – while these products sell us fun and connection, they also profit from our time and attention. This might seem like a lot to teach a five-year-old, but we felt a strong need to equip her with the ability to make decisions about her screen usage and enforce her *own* rules.

It was her *job to know when to stop because she couldn't rely upon the app makers or her parents to tell her when she'd had enough.*

We then asked her how much screen time per day she thought was good for her. We took a risk by giving her the autonomy to make the decision for herself, but it was worth a shot.

Truthfully, I expected her to say 'All day!' but she didn't. Instead, armed with the logic behind why limiting screen time was important and with the freedom to decide in her hands, she sheepishly asked for 'two shows'. Two episodes of a children's programme on Netflix is about forty-five minutes, I explained. 'Does forty-five minutes seem like the right amount of screen time per day for you?' I asked in all sincerity. She nodded in agreement and I could tell by the hint of a smile that she felt she had got the better end of the deal.

As far as I was concerned, forty-five minutes was fine, as it left plenty of time for other activities. 'How do you plan to make sure you don't watch for more than forty-five minutes per day?' I asked. Not wanting to lose the negotiation that she clearly felt she was winning, she proposed using a kitchen timer she could set herself. 'Sounds good,' I agreed. 'But if Mommy and Daddy notice you're not able to keep the promise you made to yourself and to us, we'll have to revisit this discussion,' I said, and she agreed.

This is an example of how even young children can learn to use a precommitment. Today, as a spirited ten-year-old, my daughter is still in charge of her screen time. She's made some adjustments to her self-imposed guidelines as she's grown, such as trading daily episodes for a weekend movie night. She's also replaced the kitchen timer with other tools; she now calls out to Amazon's Alexa to set a timer to let her know when she's reached her limit. The important thing is that these are her rules, not ours, and that she's in charge of enforcing them. Best of all, when her time is up it's not her dad who has to be the bad guy; the device tells her she's had enough. Without realising it, she entered into an effort pact, as described in Part 4.

Many parents want to know if there is a correct amount of time kids should be allowed to spend on their screens, but no such absolute number exists. There are too many factors at play, including the child's specific needs, what the child is doing online and the activities that screen time is replacing. The most important thing is to involve the child in the conversation and help them set their own rules. When parents impose limits without their kids' input, they are setting them up to be resentful and incentivising them to cheat the system.

It's only when kids can monitor their own behaviour that they learn the skills they need to be indistractable – even when their parents aren't around.

These strategies are no guarantee of parent–child domestic harmony. In fact, we should expect to have heated discussions about the role technology plays in our homes and in our kids' lives, just as many families have fiery debates over giving the car keys to their teenagers on a Saturday night. Discussions and, at times, respectful disagreements are a sign of a healthy family.

If there is one lesson to take away from this section, and perhaps from this entire book, it's that distraction is a problem like any other. Whether in a large corporation or in a small family, when we discuss our problems openly and in an environment where we feel safe and supported, we can resolve them together.

While it's important our kids are aware that products are designed to be highly engaging, we also need to reinforce their belief in their power to overcome distraction. It's their responsibility, as well as their right, to use their time wisely.

★ REMEMBER THIS:

- **Don't underestimate your child's ability to precommit and follow through.** Even young children can learn to use precommitments as long as they set the rules and know how to use a timer or some other binding system.
- **Consumer scepticism is healthy.** Understanding that companies are motivated to keep kids spending time watching or playing is an important part of teaching media literacy.
- **Put the kids in charge.** It's only when kids practise monitoring their own behaviour that they learn how to manage their own time and attention.

Part 7

How to Have Indistractable Relationships

34

Spread Social Antibodies Among Friends

When we are with friends we're never really alone in their company; our phones are almost assuredly present and ready to interrupt us with a poorly timed notification. Who hasn't witnessed a friend divert their attention, mid-conversation, to reflexively check their phone? Most of us simply accept these interruptions, sighing them away as a sign of the times.

Unfortunately, distraction is contagious. When smokers get together, the first one to take out their pack sends a cue and, when others notice, they do the same. In a similar way, digital devices can prompt others' behaviours. When one person takes out their phone at dinner, it acts as an external trigger. Soon, others are lost in their screens, at the expense of the conversation.

Psychologists call this phenomenon 'social contagion', and researchers have found that it influences our behaviours, from drug use to overeating.[1] It's hard to watch your weight if your spouse and kids insist on mowing down a dozen frosted doughnuts as you pick at your kale salad[2] and it's difficult to change your tech habits when your family and friends shun you in favour of their screens.

Given the enormous influence others have on our actions, how can we manage distraction around those with whom we want to spend uninterrupted quality time? How do we change our tendencies towards distraction when those around us haven't changed theirs?

The essayist and investor Paul Graham writes that societies tend to develop 'social antibodies' – defences against new harmful behaviours.[3] Consider that in 1965 42.4 per cent of adult Americans smoked, according to the Centers for Disease Control and Prevention, a number that is expected to fall to just 12 per cent by 2020.[4] Of course, legal restrictions played an important role in the precipitous decline in smoking rates. However, laws do not prevent people from smoking in their own homes, and yet that habit changed even in the absence of regulation.

I remember my parents placing ashtrays around the house in my childhood, despite being non-smokers. At the time, people smoked indoors, around children, at the office – wherever they pleased. My mother did her best to discourage smoking by providing an ashtray shaped like a bony skeleton hand, but that not so subtle reminder of the consequences of smoking was all she felt comfortable doing. In those days, it was considered strange, if not rude, to ask someone to smoke outdoors.

Today, however, things are very different. I've never owned an ashtray. No one has ever asked if they could smoke in my home; they already know the answer. It scares me to imagine the look on my wife's face if someone were to light up on our living room couch. That person wouldn't be in our house or our circle of friends for long.

How did the norms around smoking change so dramatically in just a single generation? According to Graham's theory, people adopted social antibodies to protect themselves, similar to the way our bodies fight back against bacteria and viruses that can harm us. The remedy for distraction in social situations involves the development of new norms that make it taboo to check one's phone when in the company of others.

Social norms are changing, but whether or not they change for the better is up to us.

The only way to make sure certain unhealthy behaviours are no longer acceptable is to call them out and address them with social antibodies that block their spread. This tactic worked with smoking, and it can work with digital distractions.

Let's imagine you're at a dinner party when someone takes out his phone and starts to tap away. While you probably already know that spending time on a device in an intimate social setting is rude, there's often at least one person who hasn't learned the new social norm. Embarrassing them in front of others isn't a good idea, assuming you want to remain friends; a more subtle tactic is required.

To help keep things cordial, a simple and effective approach is to ask a direct question that can snap the offender out of the phone zone by giving him two simple options: 1) excuse himself to attend to the crisis happening on his device, or 2) kindly put away his phone. The question goes like this: 'I see you're on your phone. Is everything OK?'

Remember to be sincere – after all, there might really be an emergency. But, more often than not, he'll mutter a little excuse, tuck his phone back into his pocket and start enjoying the evening again. Victory is yours! You've succeeded in tactfully spreading the social antibody against 'phubbing', a word coined by the ad agency McCann for the Macquarie Dictionary.[5]

'Phubbing', a combination of 'phone' and 'snubbing', means 'to ignore (a person or one's surroundings) when in a social situation by busying oneself with a phone or other mobile device'. The dictionary assembled experts to create the word in order to give people a way to call out the problem. Now it's up to us to start using the term so that it may become another positive social antibody in our arsenal against distractions in social settings.

Modern technologies like smartphones, tablets and laptops aren't the only sources of distraction in social situations.

Many restaurants have wall-to-wall television sets, each with a different channel flashing headline news or a sports game that can easily disrupt conversations. Because of our acceptance of having televisions playing in the background in social settings, they can be equally, if not more, pernicious at distracting our attention away from the people we're with.

Distraction among friends can take on other forms, including our own children. For example, during a recent get-together, just as a good friend began to share his personal and professional struggles, one of his children came to the table and demanded a juice box. The conversation immediately shifted to the needs of the child.

Such an innocent interruption has the ability to derail an important and sensitive conversation – the kind that solidifies close friendships. The next time we had dinner together, we made sure to put everything the kids would need, including food and drinks, in another room. The kids received clear instructions not to interrupt the adults unless someone was bleeding.

All external triggers – whether coming from our phones or our kids – deserve scrutiny to determine whether they are serving us. Our children are also better served when they learn to take care of themselves, and, by watching their parents model fellowship, they learn the importance of tuning out distractions to focus on friends. If we are not intentional about making the time and space for distraction-free discussions, we risk losing the opportunity to truly know others and allow them to truly know us.

In the same way society reduced social smoking with social antibodies, we can reduce distraction while with friends. By getting agreement from our friends and families to manage distraction and taking steps to remove external triggers that don't serve us, we can quarantine the social contagion of distraction while with the people we love.

★ REMEMBER THIS:

- **Distraction in social situations can keep us from being fully present with important people in our lives.** Interruptions degrade our ability to form close social bonds.
- **Block the spread of unhealthy behaviours.** 'Social antibodies' are ways groups protect themselves from harmful behaviours by making them taboo.
- **Develop new social norms.** We can tackle distraction among friends the same way we beat social smoking, by making it unacceptable to use devices in social situations. Prepare a few tactful phrases, such as asking 'Is everything OK?' to discourage phone usage among friends.

35

Be an Indistractable Lover

Every night my wife and I engaged in the same routine: she put our daughter to bed, brushed her teeth and freshened up. Slipping under the covers, we exchanged glances and knew it was time to do what comes naturally for a couple in bed – she began to fondle her cell phone, while I tenderly stroked the screen of my iPad. Ooh, it felt so good.

We were having a love affair with our gadgets. Apparently we weren't the only ones substituting Facebook for foreplay. According to one survey, 'almost a third of Americans would rather give up sex for a year than part with their mobile phone for that long'.[1]

Before we learned to become indistractable, the allure of notifications on our cell phones proved hard to resist. Promising to reply to just one more email after dinner quickly turned into forty-five minutes of lost intimacy later that night. We'd fallen into an evening ritual of solitary tech checking until midnight. By the time each of us got to bed, we were too tired to talk. Our relationship, not to mention our sex life, suffered.

We were among the 65 per cent of American adults who, according to the Pew Research Center, sleep with their phones on or next to their beds.[2] Since habits rely on a cue to trigger a behaviour, action is often sparked by the things around us. We decided to move our phones from our bedroom to the living room, and, with the external triggers gone, we regained a bit more control over our techno-infidelity.

But after a few phone-free evenings I began to notice a stressful anxiety. My mind became occupied with all the things calling for my attention. Had someone sent me an urgent email? What was the latest comment on my blog about? Did I miss something important on Twitter? The stress was palpable and painful, so I did what anyone who makes a firm commitment to breaking a bad habit would do: I cheated.

With my cell phone unavailable, I needed to find a new partner. To my relief, I felt the anxiety melt away as I pulled out my laptop and began to bang on the keyboard. My wife, seeing what I was doing, pounced on the opportunity to relieve her own stress, and we were back at it again.

After a few late nights on our machines, we sheepishly admitted that we had failed. Embarrassed but determined to understand where we'd gone wrong, we realised we had skipped a critical step. We hadn't learned to deal with the discomfort that had drawn us back in. With self-compassion, this time, we decided to start by finding ways to manage the internal triggers driving our unwanted behaviours.

We implemented a ten-minute rule and promised that if we really wanted to use a device in the evening we would wait ten minutes before doing so. The rule allowed us time to 'surf the urge' and insert a pause to interrupt the otherwise mindless habit.

We also connected our internet router and monitors to $7 timer outlets purchased at a local hardware store and set them to turn off at ten o'clock every evening. Using this effort pact meant that, in order to 'cheat', we would have to uncomfortably contort behind our desks and flip the override switch.

In short, we were making progress by using all four methods for becoming indistractable. We learned to cope with the stress of stopping our compulsion to use technology in the evening, and, over time, it became easier to resist. We scheduled a strict bedtime, claiming the bedroom as a sacred space and leaving external triggers, like our cell phones and the television, outside. The outlet timers that turned off the unwanted distractions made compliance with our precommitment something we came to expect every night. We began to use our reclaimed time for more 'productive' purposes as we gained greater control over our habits.

Though we were proud of our tech-blocking invention, many routers like the Eero now come with internet shut-off capabilities built in.[3] If I lose track of time and try to check email after ten o'clock, a message from my router reminds me to get off the computer and go snuggle with my wife.

Distractions can take a toll on even our most intimate relationships; the cost of being able to connect with anyone in the world is that we might not be fully present with the person physically next to us.

My wife and I still love our gadgets and fully embrace the potential of innovation to improve our lives, but we want to benefit from technology without suffering from the corrosive effects it can have on our relationship. By learning to deal with our internal triggers, making time for the things we really want to do, removing harmful external triggers, and using precommitments, we were finally able to conquer distractions in our relationship.

***

As you read in Part 1, 'Being indistractable means striving to do what you say you will do.' To strive means 'to struggle or fight vigorously',[4] it does not mean being perfect or never failing. Like everyone, I still struggle with distraction at times. When I'm particularly stressed or my schedule changes unexpectedly, I can fall off-track.

Distractions still happen, but now I know what to do about them so they don't *keep* happening. These techniques have allowed me to take control of my life in ways I never could before. I'm as honest with myself as I am with others, I live up to my values, I fulfil my commitments to the people I love, and I am more productive professionally than ever.

Recently, I revisited the conversation I'd had with my daughter about what superpower she'd want. After apologising to her for not being fully present the last time we had the conversation, I asked her to tell me her answer again, and what she said blew me away: she said she wanted the power always to be kind to others.

After drying my eyes and giving her a big hug, I took some time to think about her answer. I realised that being kind was not a mystical superpower that required a magical serum – we all have the power to be kind whenever we want to be. We simply need to harness the power that's already within us.

The same is true for being indistractable. By becoming indistractable, we can set an example for others. In the workplace, we can use these tactics to transform our organisations and create a ripple effect both in and beyond our industries. At home, we can inspire our families to test these methods for themselves and to live out the lives they envisage.

We can all strive to do what we say we will do. We all have the power to be indistractable.

★ REMEMBER THIS:

- **Distraction can be an impediment in our most intimate relationships.** Instant digital connectivity can come at the expense of being fully present with those beside us.
- **Indistractable partners reclaim time for togetherness.** Following the four steps to becoming indistractable can ensure you make time for your partner.
- **Now it's your turn to become indistractable.**

DID YOU ENJOY THIS BOOK?

Congratulations and thank you for completing this book! I hope you'll put what you've read to good use.

If you have a minute, it would mean so much to me if you would review this book online. Your review goes a long way towards encouraging other people to read *Indistractable* and I'd consider it a *huge personal favour*.

Thank you in advance! Please go to:

NirAndFar.com/ReviewIndistractable

And please send any questions, comments, edits or feedback to:

NirAndFar.com/Contact

Sincerest thanks!
Nir

CHAPTER TAKEAWAYS

- Chapter 7: Reimagine the task. Turn it into play by paying 'foolish, even absurd' attention to it. Deliberately look for novelty.
- Chapter 8: Reimagine your temperament. Self-talk matters. Your willpower runs out only if you believe it does. Avoid labeling yourself as 'easily distracted' or having an 'addictive personality'.

PART 2: MAKE TIME FOR TRACTION

- Chapter 9: Turn your values into time. Timebox your day by creating a schedule template.
- Chapter 10: Schedule time for yourself. Plan the inputs and the outcome will follow.
- Chapter 11: Schedule time for important relationships. Include household responsibilities as well as time for people you love. Put regular time on your schedule for friends.
- Chapter 12: Sync your schedule with stakeholders.

PART 3: HACK BACK EXTERNAL TRIGGERS

- Chapter 13: Of each external trigger, ask: 'Is this trigger serving me, or am I serving it?' Does it lead to traction or distraction?
- Chapter 14: Defend your focus. Signal when you do not want to be interrupted.
- Chapter 15: To get fewer emails, send fewer emails. When you check email, tag each message with when it needs a reply and respond at a scheduled time.

- Chapter 16: When it comes to group chat, get in and out at scheduled times. Only involve who is necessary and don't use it to think out loud.
- Chapter 17: Make it harder to call meetings. No agenda, no meeting. Meetings are for consensus building rather than problem solving. Leave devices outside the conference room except for one laptop.
- Chapter 18: Use distracting apps on your desktop rather than your phone. Organise apps and manage notifications. Turn on 'Do Not Disturb'.
- Chapter 19: Turn off desktop notifications. Remove potential distractions from your workspace.
- Chapter 20: Save online articles in Pocket to read or listen to at a scheduled time. Use 'multichannel multitasking'.
- Chapter 21: Use browser extensions that give you the benefits of social media without all the distractions. Links to other tools are at: NirAndFar.com/Indistractable

PART 4: PREVENT DISTRACTIONS WITH PACTS

- Chapter 22: The antidote to impulsiveness is forethought. Plan ahead for when you're likely to get distracted.
- Chapter 23: Use effort pacts to make unwanted behaviours more difficult.
- Chapter 24: Use a price pact to make getting distracted expensive.
- Chapter 25: Use identity pacts as a precommitment to a self-image. Call yourself 'indistractable'.

PART 5: HOW TO MAKE YOUR WORKPLACE
INDISTRACTABLE

- Chapter 26: An 'always on' culture drives people crazy.
- Chapter 27: Tech overuse at work is a symptom of dysfunctional company culture. The root cause is a culture lacking 'psychological safety'.
- Chapter 28: To create a culture that values doing focused work, start small and find ways to facilitate an open dialogue among colleagues about the problem.

PART 6: HOW TO RAISE INDISTRACTABLE CHILDREN
(AND WHY WE ALL NEED PSYCHOLOGICAL NUTRIENTS)

- Chapter 29: Find the root causes of why children get distracted. Teach them the four-part indistractable model.
- Chapter 30: Make sure children's psychological needs are met. All people need to feel a sense of autonomy, competence and relatedness. If kids don't get their needs met in the real world, they look to fulfill them online.
- Chapter 31: Teach children to timebox their schedule. Let them make time for activities they enjoy, including time online.
- Chapter 32: Work with your children to remove unhelpful external triggers. Make sure they know how to turn off distracting triggers. And don't become a distracting external trigger yourself.
- Chapter 33: Help your kids make pacts and make sure they know managing distraction is their responsibility.

Teach them that distraction is a solvable problem and that becoming indistractable is a lifelong skill.

PART 7: HOW TO HAVE INDISTRACTABLE RELATIONSHIPS

- Chapter 34: When someone uses a device in a social setting, ask 'I see you're on your phone. Is everything okay?'
- Chapter 35: Remove devices from your bedroom and have the internet automatically turn off at a specific time.

SCHEDULE TEMPLATE

For a free online scheduling tool visit NirAndFar.com/ Indistractable.

	Monday	Tuesday	Wednesday	Thursday	Friday	Saturday	Sunday
7:00 AM							
8:00 AM							
9:00 AM							
10:00 AM							
11:00 AM							
12:00 PM							
1:00 PM							
2:00 PM							
3:00 PM							
4:00 PM							
5:00 PM							
6:00 PM							
7:00 PM							
8:00 PM							
9:00 PM							
10:00 PM							
11:00 PM							

DISTRACTION TRACKER

(see Chapter 9 for instructions)

Time	Distraction	How feeling	Internal	External	Planning Problem	Ideas
8:15	Checked news	Anxious	X			Surf the urge
9:32	Googled instead of writing	Frustrated	x			Set time goal and see if I can beat it

BOOK CLUB DISCUSSION GUIDE

Invite a few friends to discuss the topics mentioned in this book, using these questions as a guide. The ensuing conversation around productivity, habits, values, technology and triggers will be lively, thorough and interesting – and may provoke further change.

1. Throughout the book, Nir speaks about the importance of the three life domains: you, relationships and work. Often we unintentionally spend too much time in one area at the expense of others. Which life domain do you desire to improve the most, and why?

2. *Indistractable* is full of unconventional wisdom. Was there anything Nir said that changed your mind in a way that surprised you?

3. Think about the frequent distractions that prevent you from achieving traction. What are your three most common internal triggers? Then consider what are your three most common external triggers. Remember, internal triggers cue us from within, while external triggers are cues in our environment.

4. Nir describes how we can use fun and play to reimagine a seemingly boring or repetitive task. Think about something you do in your day-to-day life or work day that isn't particularly engaging. How can you reimagine the task (or add in a constraint) to make it more interesting?

5. Creating a fun jar served Nir's goal of becoming a more involved father to his young daughter. What are five activities that would be a 'must' in your fun jar?

6. Nir offers a polarising view of to-do lists and says that they're seriously flawed and don't work. Do you agree or disagree with this statement? Why?

7. Aligning your schedule with your values is essential to achieving traction. Dream up an ideal timeboxed day in your life. How would you spend your time? How would you 'turn your values into time' for yourself, for your relationships and for your work?

8. Values are not end goals; they are guidelines for our actions. What three values are most important to you?

9. Studies have shown that the modern workplace and particularly open-plan offices are a constant source of distraction. Do you agree or disagree?

10. Distraction is inevitable at work, even when you work from home. Everything from group chats to our phones can take us off course. How will you make uninterrupted work a priority in your daily grind?

11. We learned in the book that our identities are not fixed. Like habits, we can choose to change our identities and commit to a more positive self-image. What are a few habits you've long desired to change, and how could you create a new identity to empower yourself to success?

12. Nir wrote, 'Limitations give us structure, while nothingness torments us with the tyranny of choice.' Describe an instance in your life where constraints could offer structure in a positive way.

13. Changing your behaviour is hard and people inevitably fail; it's critical to know how to bounce back from this. How have you recovered from failure in the past?

14. The internet (including social media) can be a content vortex. What habits would you like to cultivate to improve your relationship with content consumption online?

15. Nir shared an extensive list of some of his favourite hacks to combat online distraction (such as eliminating his Facebook news feed and using productivity apps like Forest). Share a hack that has helped you be more efficient and focused.

16. According to researchers, we need three psychological nutrients in order to flourish: autonomy, competence and relatedness. Which of these nutrients is most important to you and why? Which are you lacking?

17. Technological advances tend to create fear and panic (think self-driving cars, AI, virtual reality, even social media). Why do you think this is?

18. What is something you consistently fail to show up for? Maybe it's going to the gym or following up with friends. What can you do differently to make sure you do as you say by following the four parts of the Indistractable Model?

19. According to a survey, a third of Americans would give up sex for a year rather than part with their phones for that long. Which option would you give up for a year and why?

20. What is your definition of living an indistractable life?

ACKNOWLEDGEMENTS

Indistractable took over five years to complete and there were many individuals who deserve thanks for their contributions to this project.

First and foremost, my deepest thanks to my business and life partner, Julie Li. Her contributions to this project are beyond measure. Julie allowed me to share intimate stories about our marriage, was there to help me test ideas and tactics, and spent countless hours improving this book. We've walked this path together every step of the way.

Next, thank you to Jasmine, my daughter, who not only provided the inspiration to become indistractable, but who (in her ten-year-old way) was also enthusiastically helpful with the naming, cover design and marketing of the book.

And, of course, my parents, Ronit and Victor, and in-laws, Anne and Paul, for their encouragement. Their support and enthusiasm for every one of my crazy projects means so much to me.

Thank you to the brave people who read very early (and very rough) drafts of this book. Thank you to Eric Barker, Caitlin Bauer, Gaia Bernstein, Jonathan Bolden, Cara Cannella, Linda Cyr, Geraldine DeRuiter, Kyle Eschenroeder, Omer Eyal, Monique Eyal, Rand Fishkin, Jose Hamilton, Wes Kao, Josh Kaufman, Carey Kolaja, Carl Marci, Jason Ogle, Ross Overline, Taylor Pearson, Jillian Richardson, Alexandra Samuel, Oren

Shapira, Vikas Singhal, Shane Snow, Charles Wang and Andrew Zimmermann. Reading an early manuscript is hard work and I can't thank you enough for your thoughtful comments and insights.

Thank you to Christy Fletcher and her team for top-notch representation. Christy is a terrific agent and I owe her loads of thanks for her counsel and friendship. Thank you to Melissa Chinchillo, Grainne Fox, Sarah Fuentes, Veronica Goldstein, Elizabeth Resnick and Alyssa Taylor at Fletcher & Co.

I'd also like to thank Stacy Creamer at Audible, as well as the team at BenBella, including Sarah Avinger, Heather Butterfield, Jennifer Canzoneri, Lise Engel, Stephanie Gorton, Aida Herrera, Alicia Kania, Adrienne Lang, Monica Lowry, Vy Tran, Susan Welte, Leah Wilson and Glenn Yeffeth, for their efforts in bringing this book to market.

Alexis Kirschbaum at Bloomsbury went above and beyond what any author could ask for in an editor and played a critical hand in improving this book. She and her colleagues, including Hermione Davis, Thi Dinh, Genevieve Nelsson, Andy Palmer, Genista Tate-Alexander and Angelique Tran Van Sang, deserve my sincere gratitude.

Thanks to the following people for their help researching, editing and refining *Indistractable*: Karen Beattie, Matthew Gartland, Jonah Lehrer, Janna Marlies Maron, Mickayla Mazutinec, Paulette Perhach, Chelsea Robertson, Ray Sylvester and AnneMarie Ward.

Particular thanks to Thomas Kjemperud and Andrea Schumann for their assistance running NirAndFar.com. Thank you also to Carla Cruttenden and Damon Nofar for the graphics in this book, and to Rafael Arizaga Vaca for helping with more projects than I can count. I can't thank these wonderful people enough!

Then there are the following people who I want to thank for providing moral and intellectual support: Arianna Huffington, for her enthusiasm for this project; Mark Manson, Taylor Pearson and Steve Kamb, for being consistent co-working buddies and for helping me stay focused while writing this book; Adam Gazzaley, for generously providing the Indistractable. com domain; and James Clear, Ryan Holiday, David Kadavy, Fernanda Neute, Shane Parrish, Kim Raices, Gretchen Rubin, Tim Urban, Vanessa Van Edwards, Alexandra Watkins and Ryan Williams for sharing insights and giving great advice.

I have no doubt neglected to include some very important people. Along with forgiveness, I ask to invoke Hanlon's razor, 'Never attribute to malice that which is adequately explained by stupidity.' I'm sorry and thank you!

Finally, and most importantly, thank you, the reader. Spending your precious time and attention with this book means the world to me. Feel free to contact me if I can be helpful at NirAndFar.com/Contact.

CONTRIBUTORS

Thank you to the loyal blog subscribers named below for their help crowd-editing *Indistractable*. Their insights, suggestions and encouragement were immensely important in making this book what it is.

Reed Abbott

Shira Abel

Zalman Abraham

Eveline van Acquoij

Daniel Adeyemi

Patrick Adiaheno

Sachin Agarwal

Avneep Aggarwal

Vineet Aggarwal

Abhishek Kumar Agrahari

Neetu Agrawal

Sonali Agrawal

Syed Ahmed

Matteus Åkesson

Stephen Akomolafe

Alessandra Albano

Chrissy Allan

Patricia De Almeida

Hagit Alon

Bos Alvertos

Erica Amalfitano

Mateus Gundlach Ambros

Iuliia Ankudynova

Tarkan Anlar

Lauren Antonoff

Jeremi Walewicz Antonowicz

Kavita Appachu

Yasmin Aristizabal

Lara Ashmore

Aby Atilola

Jeanne Audino

Jennifer Ayers

Marcelo Schenk de Azambuja

Xavier Baars

Deepinder Singh Babbar

Rupert Bacon

Shampa Bagchi

Warren Baker

Tamar Balkin

Giacomo Barbieri

Surendra Bashani

Asya Bashina

Omri Baumer

Jeff Beckmen

Walid Belballi

Jonathan Bennun

Muna Benthami

Gael Bergeron

Abhishek Bhardwaj

Kunal Bhatia

Marc Biemer

Olia Birulia

Nancy Black

Eden Blackwell

Charlotte Blank

Kelli Blum

Rachel Bodnar

Stephan Borg

Mia Bourgeois

Charles Brewer

Sam Brinson

Jesse Brown

Michele Brown

Ryan Brown

Sarah E. Brown

Michelle E. Brownstein

John Bryan

Renée Buchanan

Scott Bundgaard

Steve Burnel

Michael Burroughs

Tamar Burton

Jessica Cameron

Jerome Cance

Jim Canterucci

Ryan Capple

Savannah Carlin

James Carman

Karla H. Carpenter

Margarida Carvalho

Anthony Catanese

Shubha Chakravarthy

Karthy Chandra

Joseph Chang

Jay Chaplin

David Chau

Janet Y. Chen

Ari Cheskes

Dennis Chirwa

Kristina Yuh-Wen Chou

Ingrid Choy-Harris

Michelle M. Chu

William Chu

Jay Chung

Matthew Cinelli

Sergiu Vlad Ciurescu

Trevor Claiborne

Kay Krystal Clopton

Heather Cloward

Lilia M. Coburn

Pip Cody
Michele Helene Cohen
Luis Colin
Abi Collins
Dave Cooper
Kerry Cooper
Simon Coxon
Carla Cruttenden
Dmitrii Cucleschin
Patrick Cullen
Leo Cunningham
Gennaro Cuofano
Ed Cutshaw
Larry Czerwonka
Lloyd D'Silva
Jonathan Dadone
Sharon F. Danzger
Kyle Huff David
Lulu Davies
James Davis, Jr.
Joel Davis
Cameron Deemer
Stephen Delaney
Keval D. Desai
Ankit S. Dhingra
Manuel Dianese
Jorge Dieguez
Lisa Hendry Dillon
Sam Dix
Lindsay Donaire

Ingrid Elise Dorai-Rekaa
Tom Droste
Nan Duangnapa
Scott Dunlap
Akhilesh Reddy Dwarampudi
Swapnil Dwivedi
Daniel Edman
Anders Eidergard
Dudi Einey
Max Elander
Ori Elisar
Katie Elliott
Gary Engel
David Ensor
Eszter Erdelyi
Ozge Ergen
Bec Evans
David Evans
Shirley Evans
Jeff Evernham
Kimberly Fandino
Kathlyn Farrell
Hannah Farrow
Michael Ferguson
Nissanka Fernando
Margaret Fero
Kyra Fillmore
Yegor Filonov
Fabian Fischer
Jai Flicker

Collin Flotta

Michael Flynn

Kaleigh Flynn

Gio Focaraccio

Ivan Foong

Michael A. Foster II

Martin Foster

Jonathan Freedman

Heather Friedland

Janine Fusco

Pooja V. Gaikwad

Mario Alberto Galindo

Mary Gallotta

Zander Galloway

Sandra Gannon

Angelica Garcia

Anyssa Sebia Garza

Allegra Gee

Tom Gilheany

Raji Gill

Scott Gillespie

Scott Gilly

Wendell Gingerich

Kevin Glynn

Paula Godar

Jeroen Goddijn

Anthony Gold

Dan Goldman

Miguel H. Gonzalez

Sandra Catalina González

Vijay Gopalakrishnan

Herve Le Gouguec

Nicholas Gracilla

Charlie Graham

Timothy L. Graham

Shawn Green

Chris Greene

Jennifer Griffin

Dani Grodsky

Rebecca Groner

Saksham Grover

Alcide Guillory III

Roberta Guise

Anjana Gummadivalli

Matt Gummow

Amit Gupta

John Haggerty

Martin Haiek

Lance Haley

Thomas Hallgren

Eric Hamilton

Caroline Hane-Weijman

Nickie Harber-Frankart

Julie Harris

Sophie Hart

Daniel Hegman

Christopher Heiser

Lisa Helminiak

Alecia Helton

Mauricio Hess-Flores

Holly Hester-Reilly

Andrea Hill

Neeraj Hirani

Isabella Catarina Hirt

Charlotte Jane Ho

Ian Hoch

Travis Hodges

Jason Hoenich

Alex J. Holte

Abi Hough

Mary Howland

Evan Huggins

Nathan Hull

Novianta L. T. Hutagalung

Marc Inzelstein

Varun Iyer

Britni Jackson

Mahaveer Jain

Abdellah Janid

Anne Janzer

Emilio Jéldrez

Debbie Jenkins

Alexandre Jeong

Amy M. Jones

Daniela Jones

Peter Jotanovic

Cindy Joung

Sarah Jukes

Steve Jungmann

Rocel Ann Junio

Kevin Just

Ahsan Kabir

Ariel Kahan

Sina Kahen

Sarah Kajani

Angela Kapdan

Shaheen Karodia

Irene Jena Karthik

Melissa Kaufmann

Gagandeep Kaur

Megan Keane

J. Bavani Kehoe

Karen Kelvie

Erik Kemper

Raye Keslensky

Jenny Shaw Kessler

Jeremy C. Kester

Kirk Ketefian

Nathan Khakshouri

Sarah Khalid

Sam Kirk

Rachel Kirton

Vinod Kizhakke

Samuel Koch

Alaina Koerber

Sai Prabhu Konchada

Jason Koprowski

Basavaraj Koti

Yannis Koutavas

David Kozisek

Aditya Kshirsagar

Ezekiel Kuang

Craig Kulyk

Ram Kunda

Ravi Kurani

Chris Kurdziel

Dimitry Kushelevsky

John Kvasnic

Jonathan Lai

Michael J. Lally

Roy Lamphier

Craig Lancaster

Niklas Laninge

Simon Lapscher

Angelo Larocca

Norman Law

Olga Lefter

Tory Leggat

Ieva Lekaviciute

Audrey Leung

Viviana Leveghi

Isaac E. H. Lewis

Belly Li

Sammy Chen Li

Philip Li

Robert Liebert

Brendan Lim

Carissa Lintao

Ross Lloyd Lipschitz

Mitchell Lisle

Mike Sho Liu

Shelly Eisen Livneh

John Loftus

Philip K. Lohr

Sune Lomholt

Sean Long

Alexis Longinotti

Glen Lubbert

Ana Lugard

Kenda Macdonald

Boykie Mackay

Andy Maes

Kristof Maeyens

Lisa Maldonado

Amin Malik

Danielle Manello

Frank Manue, Jr.

Dan Mark

Kendra Markle

Ben Marland

Rob Marois

Judy Marshall

Levi Mårten

Denise J. Martin

Megan Martin

Kristina Corzine Martinez

Saji Maruthurkkara

Laurent Mascherpa

Mark Mavroudis

Ronny Max

Eva A. May

Lisa McCormack

Gary McCue

Michael McGee

Robert McGovern

Lyle McKeany

Sarah McKee

Marisa McKently

Erik van Mechelen

Hoda Mehr

Jonathan Melhuish

Sheetal G. Melwani

Ketriel J. Mendy

Valerae Mercury

Andreia Mesquita

Johan Meyer

Kaustubh S. Mhatre

Stéphanie Michaux

Ivory Miller

Jason Ming

Al Ming

Jan Miofsky

Ahmed A. Mirza

Peter Mitchell

Mika Mitoko

Meliza Mitra

Subarna Mitra

Aditya Morarka

Amina Moreau

David Morgan

Renee F. Morris

Matthew Morrisson

Alexandra Moxin

Alex Moy

Brian Muldowney

Namrata Mundhra

Jake Munsey

Mihnea Munteanu

Kevin C. Murray

Serdar Muslu

Karan Naik

Isabelle Di Nallo

Jeroen Nas

Vaishakhi Nayar

Jordan Naylor

Christine Neff

Jamie Nelson

Kemar Newell

Lewis Kang'ethe Ngugi

Chi Gia Nguyen

Christopher Nheu

Gerard Nielsen

Adam Noall

Tim Noetzel

Jason Nokes

Craig Norman

Chris Novell

Thomas O'Duffy

Scott Oakes

Cheily Ochoa

Leon Odey-Knight
Kelechi Okorie
Oluwatobi Oladiran
Valary Oleinik
Sue Olsen
Alan Olson
Gwendolyn Olton
Maaike Ono-Boots
Brian Ostergaard
Roland Osvath
Renz Pacheco
Nina Pacifico
Sumit Pahwa
Girri M. Palaniyapan
Vishal Kumar Pallerla
Rohit Pant
Chris V. Papadimitriou
Nick Pape
Divya Parekh
Rich Paret
Alicia Park
Aaron Parker
Steve Parkinson
Mizue Parrott
Lomit Patel
Manish Patel
Swati Patil
Jon Pederson
Alon Peled
Rodaan Peralta-Rabang

Marco Perlman
Christina Diem Pham
Hung Phan
Ana Pischl
Keshav Pitani
Rose La Prairie
Indira Pranabudi
Anne Curi Preisig
Julie Price
Martin Pritchard
Rungsun Promprasith
Krzysztof Przybylski
Edmundas Pučkorius
Călin Pupăză
Daisy Qin
Lien Quach
Colin Raab
Kelly Ragle
Ruta Raju
Lalit Raju
Kim Ramirez
Prashanthi Ravanavarapu
Gustavo Razzetti
Omar Regalado
Scott W. Rencher
Brian Rensing
Joel Rigler
Gina Riley
Michelle Riley
Ioana Rill

Mark Rimkus
Cinzia Rinelli
Bridgitt Ann Robertson
Chelsea Lyn Robertson
Reigh Robitaille
Annette Rodriguez
Cynthia Rodriguez
Charles François Roels
Linda Rolf
Edgar Roman
Mathieu Romary
Jamie Rosen
Al Rosenberg
Joy Rosenstein
Christian Röß
Megan Rounds
Ruzanna Rozman
Isabel Russ
Mark Ruthman
Alex Ryan
Kimberly Ryan
Samantha Ryan
Jan Saarmann
Guy Saban
Victoria Sakal
Luis Saldana
Daniel Tarrago Salengue
Gabriel Michael Salim
Jessica Salisbury
Rick Salsa

Francesco Sanavio
Antonio J. Martinez Sanchez
Moses Sangobiyi
Julia Saxena
Stephanie Schiller
Lynnsey Schneider
Kirk Schueler
Katherine Schuetzner
Jon Seaton
Addy Suhairi Selamat
Vishal Shah
Keshav Sharma
Ruchil Sharma
Shashi Sharma
Ashley Sheinwald
Stephanie Sher
Jing Han Shiau
Claire Shields
Greg Shove
Karen Shue
Kome Sideso
David Marc Siegel
Dan Silberberg
Bianca Silva
Brian L. Silva
Mindy Silva
Zach Simon
Raymond Sims
Shiv Sivaguru
Malin Sjöstrand

Antoine Smets

Sarah Soha

Steven Sohcot

Kaisa Soininen

David Spencer

James Taylor Stables

Kurt Stangl

Laurel Stanley

John A. Starmer

Juliano Statdlober

Christin Staubo

Ihor Stecko

Nick Di Stefano

Murray Steinman

Alexander Stempel

Seth Sternberg

Anthony Sterns

Shelby Stewart

Adam Stoltz

Alan Stout

Carmela Stricklett

Scott Stroud

Swetha Suresh

Sarah Surrette

Cathleen Swallow

Bryan Sykes

Eric Szulc

Lilla Tagai

Michel Tagami

J. P. Tanner

Shantanu Tarey

Claire Tatro

Harry E. Tawil

Noreen Teoh

C. J. Terral

Amanda Tersigni

Matt Tharp

Nay Thein

Brenton Thornicroft

Julianne Tillmann

Edwin Tin

Avegail Tizon

Zak Tomich

Roger Toor

Anders Toxboe

Jimmy Tran

Tom Trebes

Artem Troinoi

Justin Trugman

Kacy Turelli

Kunal Haresh Udani

Christian von Uffel

Jason Ugie

Matt Ulrich

Branislav Vajagić

Lionel Zivan Valdellon

Steve Valiquette

Jared Vallejo

René Van der Veer

Anulekha Venkatram

Poornima Vijayashanker

Claire Viskovic

Brigit Vucic

Thuy Vuong

Sean Wachsman

Maurizio Wagenhaus

Amelia Bland Waller

Shelley Walsh

Trish Ward

Levi Warvel

Kafi Waters

Adam Waxman

Jennifer Wei

Robin Tim Weis

Patrick Wells

Gabriel Werlich

Scott Wheelwright

Ed Wieczorek

Ward van de Wiel

Hannah Mary Williams

Robert Williger

Jean Gaddy Wilson

Rob Wilson

Claire Winter

Trevor Witt

Fanny Wu

Alex Wykoff

Maria Xenidou

Raj Yadav

Josephine Yap

Arsalan Yarveisi

Yoav Yechiam

Andrew Yee

Paul Anthony Yu

Mohamad Izwan Zakaria

Jeannie Zapanta

Anna Zaremba

Renee Zau

Ari Zelmanow

Linda Zespy

Fei Zheng

Rona Zhou

Lotte Zwijnenburg

NOTES

INTRODUCTION

1 'Amazon Best Sellers: Best Industrial Product Design' (Accessed 29 October 2017), https://www.amazon.com/gp/bestsellers/books/7921653011/ref=pd_zg_hrsr_b_1_6_last.

2 Paul Virilio, *Politics of the Very Worst*, New York: Semiotext(e), 1999, p. 89.

CHAPTER 1

1 A play on a Marthe Troly-Curtin quote, 'Time You Enjoy Wasting Is Not Wasted Time', *Quote Investigator* (Accessed 19 August 2018), https://quoteinvestigator.com/2010/06/11/time-you-enjoy/.

CHAPTER 2

1 Euripides, *Orestes*, 4–13.

2 August Theodor Kaselowsky, *Tantalus and Sisyphus in Hades*, oil painting (c. 1850, now destroyed), in the Niobidensaal of the Neues Museum, Berlin, Germany, nineteenth century, https://commons.wikimedia.org/wiki/File:Tantalus-and-sisyphus-in-hades-august-theodor-kaselowsky.jpg

3 Online Etymology Dictionary, s.v. 'distraction' (Accessed 15 January 2018), https://www.etymonline.com/word/distraction.

4 Louis Anslow, 'What Technology Are We Addicted to This Time?' *Timeline* (27 May 2016), https://timeline.com/what-technology-are-we-addicted-to-this-time-f0f7860f2fab#.rfzxtvjil.

5 Plato, *Phaedrus*, trans. Benjamin Towett, 277a3–4, http://classics.mit.edu/Plato/phaedrus.html.

6   H. A. Simon, 'Designing Organizations for an Information-Rich World' in Martin Greenberger (ed.), *Computers, Communication, and the Public Interest*, Baltimore, MD: Johns Hopkins Press, 1971, pp. 40–41.

7   Hikaru Takeuchi et al., 'Failing to Deactivate: The Association between Brain Activity during a Working Memory Task and Creativity', *NeuroImage* 55, no. 2 (15 March 2011): 681–87, https://doi.org/10.1016/j.neuroimage.2010.11.052; Nelson Cowan, 'The Focus of Attention As Observed in Visual Working Memory Tasks: Making Sense of Competing Claims', *Neuropsychologia* 49, no. 6 (May 2011): 1401–6, https://doi.org/10.1016/j.neuropsychologia.2011.01.035; P. A. Howard-Jones and S. Murray, 'Ideational Productivity, Focus Of Attention, and Context', *Creativity Research Journal* 15, no. 2–3 (2003): 153–66, doi:10.1080/10400419.2003.9651409; Nilli Lavie, 'Distracted and Confused?: Selective Attention under Load', *Trends in Cognitive Sciences* 9, no. 2 (1 February 2005): 75–82, https://doi.org/10.1016/j.tics.2004.12.004; Barbara J. Grosz and Peter C. Gordon, 'Conceptions of Limited Attention and Discourse Focus', *Computational Linguistics* 25, no. 4 (1999): 617–24, http://aclweb.org/anthology/J/J99/J99-4006; Amanda L. Gilchrist and Nelson Cowan, 'Can the Focus of Attention Accommodate Multiple, Separate Items?', *Journal of Experimental Psychology, Learning, Memory, and Cognition* 37, no. 6 (November 2011): 1484–1502, https://doi.org/10.1037/a0024352.

8   Julianne Holt-Lunstad, Timothy B. Smith and J. Bradley Layton, 'Social Relationships and Mortality Risk: A Meta-Analytic Review', *PLOS Medicine* 7, no. 7 (27 July 2010): e1000316. https://doi.org/10.1371/journal.pmed.1000316.

CHAPTER 3

1   Zoë Chance, 'How to Make a Behavior Addictive', TEDx talk at TEDxMillRiver (14 May 2013), http://www.youtube.com/watch?v=AHfiKav9fcQ.

2   Zoë Chance in interview with the author on 16 May 2014.

3 Jeremy Bentham, *An Introduction to the Principles of Morals and Legislation*, new edition, corrected by the author (1823; repr., Oxford: Clarendon Press, 1907), http://www.econlib.org/library/Bentham/bnthPML1.html.

4 Epicurus, 'Letter to Menoeceus', contained in *Diogenes Laertius, Lives of Eminent Philosophers,* Book X.

5 Paul F. Wilson, Larry D. Dell and Gaylord F. Anderson, *Root Cause Analysis: A Tool for Total Quality Management* (Milwaukee, WI: American Society for Quality, 1993).

6 Zoë Chance in email exchange with author, 11 July 2014.

CHAPTER 4

1 Max Roser, 'The Short History of Global Living Conditions and Why It Matters That We Know it, *Our World in Data* (Accessed 30 December 2017), https://ourworldindata.org./a-history-of-global-living-conditions-in-5-charts.

2 Adam Gopnik, 'Man of Fetters', *New Yorker* (1 December 2008), https://www.newyorker.com/magazine/2008/12/08/man-of-fetters.

3 R. F. Baumeister et al., 'Bad Is Stronger Than Good', *Review of General Psychology* 5, no. 4 (December 2001): 323–70, https://doi.org/10.1037/1089-2680.5.4.323.

4 Timothy D. Wilson et al., 'Just Think: The Challenges of the Disengaged Mind', *Science* 345, no. 6192 (4 July 2014), 75–7. https://doi.org/10.1126/science.1250830.

5 'Top Sites in United States', *Alexa* (Accessed 30 December 2017), http://www.alexa.com/topsites/countries/US.

6 Jing Chai et al., 'Negativity Bias in Dangerous Drivers', *PLOS ONE* 11, no. 1 (14 January 2016), e0147083, https://doi.org/10.1371/journal.pone.0147083.

7 R. F. Baumeister et al., 'Bad Is Stronger Than Good', op. cit.

8 A. Vaish, T. Grossmann and A. Woodward, 'Not all emotions are created equal: The negativity bias in social-emotional development', *Psychological Bulletin* 134, no. 3 (2008): 383–403, https://doi.org/10.1037/0033-2909.134.3.383.

9 R. F. Baumeister et al., 'Bad Is Stronger Than Good', op. cit.

10 Wendy Treynor, Richard Gonzalez and Susan Nolen-Hoeksema, 'Rumination Reconsidered: A Psychometric Analysis', *Cognitive Therapy and Research* 27, no. 3 (1 June 2003): 247–59, https://doi.org/10.1023/A:1023910315561.

11 N. J. Ciarocco, K. D. Vohs and R. F. Baumeister, 'Some Good News About Rumination: Task-Focused Thinking After Failure Facilitates Performance Improvement', *Journal of Social and Clinical Psychology*, 29, no.10 (2010): 1057–73, http://assets.csom.umn.edu/assets/166704.pdf.

12 K. M. Sheldon and S. Lyubomirsky, 'The Challenge of Staying Happier: Testing the Hedonic Adaptation Prevention Model', *Personality and Social Psychology Bulletin* 38 (23 February 2012): 670, http://sonjalyubomirsky.com/wp-content/themes/sonjalyubomirsky/papers/SL2012.pdf.

13 David Myers, *The Pursuit of Happiness*, New York: William Morrow & Co., 1992, p. 53.

14 Richard E. Lucas et al., 'Reexamining Adaptation and the Set Point Model of Happiness: Reactions to Changes in Marital Status', *Journal of Personality and Social Psychology* 84(3) (2003): 527–39, http://www.apa.org/pubs/journals/releases/psp-843527.pdf.

CHAPTER 5

1 'Jonathan Bricker, Psychologist and Smoking Cessation Researcher', Fred Hutch (Accessed 4 February 2018), http://www.fredhutch.org/en/diseases/featured-researchers/bricker-jonathan.html.

2 Fyodor Dostoevsky, *Writer Notes on Summer Impressions*, trans. David Patterson (1988; repr., Evanston: Northwestern University Press, 1997).

3 Lea Winerman, 'Suppressing the "White Bears"', *Monitor on Psychology* 42, no. 9 (October 2011), http://www.apa.org/monitor/2011/10/unwanted-thoughts.aspx.

4 Nicky Blackburn, 'Smoking – A Habit Not an Addiction', *Israel21c* (18 July 2010), http://www.israel21c.org/smoking-a-habit-not-an-addiction/.

5 Reuven Dar et al., 'The Craving To Smoke In Flight Attendants: Relations With Smoking Deprivation, Anticipation Of Smoking,

And Actual Smoking', *Journal of Abnormal Psychology* 119, no. 1 (2010): 248–53, https://doi.org/10.1037/a0017778.

6 Cecilia Cheng and Angel Yee-lam Li, 'Internet Addiction Prevalence and Quality of (Real) Life: A Meta-Analysis of 31 Nations Across Seven World Regions', *Cyberpsychology, Behavior, and Social Networking* 17, no. 12 (1 December 2014): 755–60. https://doi.org/10.1089/cyber.2014.0317.

CHAPTER 6

1 Jonathan Brickes in conversation with the author, August 2017.

2 Judson A. Brewer et al., 'Mindfulness Training for Smoking Cessation: Results from a Randomized Controlled Trial', *Drug and Alcohol Dependence* 119, no. 1–2 (December 2011): 72–80, https://doi.org/10.1016/j.drugalcdep.2011.05.027.

3 Kelly McGonigal, *The Willpower Instinct: How Self-Control Works, Why It Matters, and What You Can Do to Get More of It*, New York: Avery Publishing, 2011.

4 'Riding the Wave: Using Mindfulness to Help Cope with Urge', *Portland Psychotherapy* (18 November 2011), https://portlandpsychotherapyclinic.com/2011/11/riding-wave-using-mindfulness-help-cope-urges/.

5 Sarah Bowen and Alan Marlatt, 'Surfing the Urge: Brief Mindfulness-Based Intervention for College Student Smokers', *Psychology of Addictive Behaviors: Journal of the Society of Psychologists in Addictive Behaviors* 23, no. 4 (December 2009): 666–71, https://doi.org/10.1037/a0017127.

6 Oliver Burkeman, 'If You Want to Have a Good Time, Ask a Buddhist', *The Guardian* (17 August 2018), https://www.theguardian.com/lifeandstyle/2018/aug/17/want-have-good-time-ask-a-buddhist.

CHAPTER 7

1 Ian Bogost, *Play Anything: The Pleasure of Limits, the Uses of Boredom, and the Secret of Games*, New York: Basic Books, 2016, p. 19.

2 'The Cure for Boredom Is Curiosity. There Is No Cure for Curiosity – Quote Investigator' (Accessed 4 March 2019), https://quoteinvestigator.com/2015/11/01/cure/.

CHAPTER 8

1 'Temperament | Definition of Temperament in English by Oxford Dictionaries,' *Oxford Dictionaries* (Accessed 17 August 2018), https://en.oxforddictionaries.com/definition/temperament.

2 Roy F. Baumeister and John Tierney, *Willpower: Rediscovering the Greatest Human Strength,* reprint, New York: Penguin Books, 2012.

3 M. T. Gailliot et al., 'Self-Control Relies on Glucose as a Limited Energy Source: Willpower Is More than a Metaphor', NCBI *PubMed* (Accessed 4 February 2018), https://www.ncbi.nlm.nih.gov/pubmed/17279852.

4 Evan C. Carter and Michael E. McCullough, 'Publication Bias and the Limited Strength Model of Self-Control: Has the Evidence for Ego Depletion Been Overestimated?' *Frontiers in Psychology* 5 (2014), https://doi.org/10.3389/fpsyg.2014.00823.

5 Evan C. Carter et al., 'A Series of Meta-Analytic Tests of the Depletion Effect: Self-Control Does Not Seem to Rely on a Limited Resource', *Journal of Experimental Psychology, General* 144, no. 4 (August 2015): 796–815, https://doi.org/10.1037/xge0000083.

6 Rob Kurzban, 'Glucose Is Not Willpower Fuel', *Evolutionary Psychology* blog archive (Accessed 4 February 2018), http://web.sas.upenn.edu/kurzbanepblog/2011/08/29/glucose-is-not-willpower-fuel/; Miguel A. Vadillo, Natalie Gold and Magda Osman, 'The Bitter Truth About Sugar and Willpower: The Limited Evidential Value of the Glucose Model of Ego Depletion', *Psychological Science* 27, no. 9 (1 September 2016): 1207–14, https://doi.org/10.1177/0956797616654911.

7 Veronika Job et al., 'Beliefs about Willpower Determine the Impact of Glucose on Self-Control', *Proceedings of the National Academy of Sciences* 110, no. 37 (10 September 2013): 14837–42, https://doi.org/10.1073/pnas.1313475110.

8 Michael Inzlicht, 'Research' (Accessed 4 February 2018), http://michaelinzlicht.com/research/.

9 'Craving Beliefs Questionnaire' (Accessed 17 August 2018), https://drive.google.com/a/nireyal.com/file/d/0B0Q6Jkc_9z2DaHJaTndPMVVkY1E/view?usp=drive_open&usp=embed_facebook.

10 Nicole Lee et al., 'It's the Thought That Counts: Craving Metacognitions and Their Role in Abstinence from Methamphetamine Use', *Journal of Substance Abuse Treatment* 38 (1 April 2010): 245–50, https://doi.org/10.1016/j.jsat.2009.12.006.

11 Elizabeth Nosen and Sheila R. Woody, 'Acceptance of Cravings: How Smoking Cessation Experiences Affect Craving Belief', *Behaviour Research and Therapy* 59 (August 2014): 71–81, https://doi.org/10.1016/j.brat.2014.05.003.

12 Hakan Turkcapar et al., 'Beliefs as a Predictor of Relapse in Alcohol-Dependent Turkish Men', *Journal of Studies on Alcohol and Drugs* 66, no. 6 (1 November 2005): 848–51, https://doi.org/10.15288/jsa.2005.66.848.

13 Steve Matthews, Robyn Dwyer and Anke Snoek, 'Stigma And Self-Stigma In Addiction', *Journal of Bioethical Inquiry* 14, no. 2 (2017): 275–86, https://doi.org/10.1007/s11673-017-9784-y.

14 Ulli Zessin, Oliver Dickhäuser and Sven Garbade, 'The Relationship Between Self-Compassion and Well-Being: A Meta-Analysis', *Applied Psychology, Health and Well-Being* 7, no. 3 (November 2015): 340–64, https://doi.org/10.1111/aphw.12051.

15 Denise Winterman, 'Rumination: The Danger of Dwelling', BBC News (17 October 2013), https://www.bbc.com/news/magazine-24444431.

CHAPTER 9

1 Johann Wolfgang von Goethe, *Maxims and Reflections*, London: Penguin Books, 1999.

2 Lucius Annaeus Seneca, *On the Shortness of Life*, trans C. D. N. Costa, New York: Penguin Books, 2005.

3 S. Kuruvilla, 'A Study Of Calendar Usage In The Workplace', *Static* (Accessed 31 January 2018), http://static.ppai.org/documents/business%20study%20final%20report%20version%204.pdf.

4 Nod to Zig Ziglar, who phrased it slightly differently, writing, 'If you don't plan your time, someone else will help you waste it.' Zig Ziglar and Tom Ziglar, *Born to Win: Find Your Success Code*, Seattle: Made for Success Publishing, 2012, p.52.

5   Russ Harris and Steven Hayes, *The Happiness Trap: How to Stop Struggling and Start Living: A Guide to ACT*, Boston: Trumpeter, 2008, p. 167.

6   Massimo Pigliucci, 'When I Help You, I Also Help Myself: On Being a Cosmopolitan', *Aeon* (Accessed 15 January 2018), https://aeon.co/ideas/when-i-help-you-i-also-help-myself-on-being-a-cosmopolitan.

7   S. Kaufman, 'Does Creativity Require Constraints?' *Psychology Today* (30 August 2011), https://www.psychologytoday.com/blog/beautiful-minds/201108/does-creativity-require-constraints.

8   P. M. Gollwitzer, 'Implementation Intentions: Strong Effects of Simple Plans', *American Psychologist* 54 (1999): 493–503, https://dx.doi.org/10.1037/0003-066x.54.7.493.

CHAPTER 10

1   L. Lamberg, 'Adults Need 7 or More Hours of Sleep Every Night', *Psychiatric News* (17 September 2015), https://psychnews.psychiatryonline.org/doi/10.1176/appi.pn.2015.9b12.

2   'What Causes Insomnia?' National Sleep Foundation (Accessed 11 September 2018), https://sleepfoundation.org/insomnia/content/what-causes-insomnia.

CHAPTER 11

1   David S. Pedulla and Sarah Thébaud, 'Can We Finish the Revolution? Gender, Work-Family Ideals, and Institutional Constraint', *American Sociological Review* 80, no. 1 (1 February 2015): 116–39, https://doi.org/10.1177/0003122414564008.

2   Darcy, Lockman, 'Analysis | Where Do Kids Learn to Undervalue Women? From Their Parents', *Washington Post*, 10 November 2017, sec. Outlook https://www.washingtonpost.com/outlook/where-do-kids-learn-to-undervalue-women-from-their-parents/2017/11/10/724518b2-c439-11e7-afe9-4f6ob5a6c4ao_story.html.

3   George E. Vaillant, Xing-jia Cui and Stephen Soldz, 'The Study of Adult Development', Harvard Department of Psychiatry (Accessed 9 November 2017), http://www.hms.harvard.edu/psych/redbook/redbook-family-adult-01.htm.
4   Robert Waldinger, 'The Good Life', TEDx talk at TEDxBeaconStreet (30 November 2015), https://www.youtube.com/ watch?v=q- 7zAkwAOYg.
5   Julie Beck, 'How Friendships Change in Adulthood', *The Atlantic* (22 October 2015), https://www.theatlantic.com/health/archive/2015/10/how-friendships-change-over-time-in-adulthood/411466/.

CHAPTER 12

1   'Neverfail Mobile Messaging Trends Study Finds 83 Percent of Users Admit to Using a Smartphone to Check Work Email After Hours', *Neverfail* via PRNewswire (22 November 2011), https://www.prnewswire.com/news-releases/neverfail-mobile-messaging-trends-study-finds-83-percent-of-users-admit-to-using-a-smartphone-to-check-work-email-after-hours-134314168.html.
2   Marianna Virtanen et al., 'Long Working Hours and Cognitive Function: The Whitehall II Study', *American Journal of Epidemiology* 169, no.5 (2008): 596–605, http://dx.doi.org/10.1093/aje/kwn382.

CHAPTER 13

1   Wendy in interviews with author, January 2018.
2   'Hack | Definition of Hack in English by Oxford Dictionaries', *Oxford Dictionaries | English* (Accessed 11 September 2018), https://en.oxforddictionaries.com/definition/hack.
3   Mike Allen, 'Sean Parker Unloads on Facebook: "God Only Knows What It's Doing to Our Children's Brains"', *Axios* (9 November 2017), https://www.axios.com/sean-parker-unloads-on-facebook-2508036343.html.
4   Edward L. Deci and Richard M. Ryan, 'Self-determination Theory: A Macrotheory of Human Motivation, Development, and

Health', *Canadian Psychology/Psychologie Canadienne* 49, no. 3 (2008): 182–5, https://doi.org/10.1037/a0012801.

5   David Pierce, 'Turn Off Your Push Notifications. All of Them', *Wired* (23 July 2017), https://www.wired.com/story/turn-off-your-push-notifications/.

6   Gloria Mark, Daniela Gudith and Ulrich Klocke, 'The Cost Of Interrupted Work: More Speed And Stress', UC Donald Bren School of Information & Computer Sciences (Accessed 20 February 2018), https://www.ics.uci.edu/~gmark/chi08-mark.pdf.

7   C. Stothart, A. Mitchum and C. Yehnert, 'The Attentional Cost of Receiving a Cell Phone Notification', *Journal of Experimental Psychology: Human Perception and Performance* 41, no. 4 (August 2015): 893–97, http://dx.doi.org/10.1037/xhp0000100.

8   Lori A. J. Scott-Sheldon et al., 'Text Messaging-Based Interventions for Smoking Cessation: A Systematic Review and Meta-Analysis', *JMIR mHealth and uHealth* 4, no. 2 (20 May 2016), https://doi.org/DOI:10.2196/mhealth.5436.

9   'Study Reveals Success of Text Messaging in Helping Smokers Quit: Text Messaging Interventions to Help Smokers Quit Should Be a Public Health Priority, Study Says', *ScienceDaily* (Accessed 27 November 2017), https://www.sciencedaily.com/releases/2016/05/160523141214.htm.

CHAPTER 14

1   Philip Aspden et al., eds, *Preventing Medication Errors: Consensus Study Report*, Institute of Medicine of the National Academies (2006), https://doi.org/10.17226/11623.

2   Maggie Fox and Lauren Dunn, 'Could Medical Errors Be the No. 3 Cause of Death?' *NBC News* (4 May 2016), https://www.nbcnews.com/health/health-care/could-medical-errors-be-no-3-cause-death-america-n568031.

3   Victoria Colliver, 'Prescription for Success: Don't Bother Nurses,' *SFGate* (28 October 2009), http://www.sfgate.com/health/article/Prescription-for-success-Don-t-bother-nurses-3282968.php.

4   Debra Wood, 'Decreasing Disruptions Reduces Medication Errors',
    RN.com (Accessed 8 December 2017), https://www.rn.com/Pages/
    ResourceDetails.aspx?id=3369.

5   Innovation Consultancy, 'Sanctifying Medication Administration',
    KP MedRite (Accessed October 2018), https://xnet.kp.org/
    innovationconsultancy/kpmedrite.html.

6   Victoria Colliver, 'Prescription for Success: Don't Bother Nurses',
    op. cit.

7   'Code of Federal Regulations: Part 121 Operating Requirements:
    Domestic, Flag, and Supplemental Operations', Federal Aviation
    Administration (Accessed December 2017), http://rgl.faa.gov/
    Regulatory_and_Guidance_Library/rgFAR.nsf/0/dd19266cebdac9
    db852566ef006d346f!OpenDocument.

8   Debra Wood, 'Decreasing Disruptions Reduces Medication Errors',
    op. cit.

9   Nick Fountain and Stacy Vanels Smith, 'Episode 704: Open Office',
    NPR's *Planet Money* (8 August 2018), https://www.npr.org/sections/
    money/2018/08/08/636668862/episode-704-open-office.

10 Yousef Alhorr et al., 'Occupant Productivity and Office Indoor
    Environment Quality: A Review of the Literature', *Building
    and Environment* 105 (1 June 2016), https://doi.org/10.1016/
    j.buildenv.2016.06.001.

11 Jeffrey Joseph, *Do Open/Collaborative Work Environments Increase,
    Decrease Or Tend To Keep Employee Satisfaction Neutral?*, ebook,
    Cornell University ILR School (2016), https://digitalcommons.ilr.
    cornell.edu/cgi/viewcontent.cgi?referer=https://www.google.ca/
    &httpsredir=1&article=1098&context=student.

CHAPTER 15

1   Sara Radicati, ed., *Email Statistics Report 2014–2018*, (2014),
    http://www.radicati.com/wp/wp-content/uploads/2014/01/Email-
    Statistics-Report-2014-2018-Executive-Summary.pdf.

2   Thomas Jackson, Ray Dawson and Darren Wilson, 'Reducing the
    Effect of Email Interruptions on Employees', *International Journal
    of Information Management* 23, no. 1 (February 2003): 55–65,
    https://doi.org/10.1016/S0268-4012(02)00068-3.

3 Michael Mankins, 'Why the French Email Law Won't Restore Work-Life Balance', *Harvard Business Review* (6 January 2017), https://hbr.org/2017/01/why-the-french-email-law-wont-restore-work-life-balance.

4 Sam McLeod, 'Skinner—Operant Conditioning', *Simply Psychology* (2018), https://www.simplypsychology.org/operant-conditioning.html.

5 'Delay or Schedule Sending Email Messages', Microsoft Office Support, https://support.office.com/en-us/article/delay-or-schedule-sending-email-messages-026af69f-c287-490a-a72f-6c65793744ba.

6 https://mixmax.com/.

7 https://www.sanebox.com/signup/6e200350ab.

8 Kostadin Kushlev and Elizabeth W. Dunn. 'Checking Email Less Frequently Reduces Stress', *Computers in Human Behavior* 43 (1 February 2015): 220–28. https://doi.org/10.1016/j.chb.2014.11.005.

CHAPTER 16

1 Jason Fried, 'Is Group Chat Making You Sweat?' *Signal v. Noise* (7 March 2016), https://m.signalvnoise.com/is-group-chat-making-you-sweat.

2 Ibid.

CHAPTER 17

1 *The Year Without Pants: wordpress.com and the Future of Work* (San Franciso: Jossey-Blass, 2017), p. 42.

2 Middlebrooks, Catherine D., Tyson Kerr, and Alan D. Castel, 'Selectively Distracted: Divided Attention and Memory for Important Information', *Psychological Science* 28, no. 8 (August 2017): 1103–15. https://doi.org/10.1177/0956797617702502; Larry Rosen and Alexandra Samuel, 'Conquering Digital Distraction', *Harvard Business Review* (1 June 2015), https://hbr.org/2015/06/conquering-digital-distraction.

CHAPTER 18

1 'Principles of Drug Addiction Treatment: A Research-Based Guide (Third Edition)', National Institute on Drug Abuse, 17 January 2018, https://www.drugabuse.gov/publications/principles-drug-addiction-treatment-research-based-guide-third-edition.

2 Tony Stubblebine, 'How to Configure Your Cell Phone for Productivity and Focus', *Better Humans* (24 August 2017), https://betterhumans.coach.me/how-to-configure-your-cell-phone-for-productivity-and-focus-1e8bd8fc9e8d.

3 David Pierce, 'Turn Off Your Push Notifications. All of Them', *Wired* (23 July 2017), https://www.wired.com/story/turn-off-your-push-notifications/.

4 Conversation with author, January 2016.

5 'How to Use Do Not Disturb While Driving', *Apple Support* (Accessed 5 December 2017), https://support.apple.com/en-us/HT208090.

CHAPTER 19

1 Stephanie McMains and Sabine Kastner, 'Interactions of Top-down and Bottom-up Mechanisms in Human Visual Cortex', *Journal of Neuroscience* 31, no. 2 (12 January 2011): 587–97, https://doi.org/10.1523/JNEUROSCI.3766-10.2011.

2 Marketta Niemelä and Pertti Saariluoma, 'Layout Attributes and Recall', *Behaviour & Information Technology* 22, no. 5 (1 September 2003): 353–63, https://doi.org/10.1080/0144929031000156924.

3 Sophie Leroy, 'Why Is It so Hard to Do My Work? The Challenge of Attention Residue When Switching between Work Tasks', *Organizational Behavior and Human Decision Processes* 109, no. 2 (1 July 2009): 168–81, https://doi.org/10.1016/j.obhdp.2009.04.002.

CHAPTER 20

1 https://getpocket.com/.

2 Claudia Wallis, 'GenM: The Multitasking Generation', *Time* (27 March 2006), http://content.time.com/time/magazine/article/0,9171,1174696,00.html.

3   B. Rapp and S. K. Hendel, 'Principles of Cross-modal Competition: Evidence from Deficits of Attention', *Psychonomic Bulletin & Review* 10, no. 1 (2003): 210–19.

4   May Wong, 'Stanford Study Finds Walking Improves Creativity', *Stanford News* (24 April 2014), https://news.stanford.edu/2014/04/24/walking-vs-sitting-042414/.

5   Katherine L. Milkman, Julia A. Minson and Kevin G. M. Volpp, 'Holding the Hunger Games Hostage at the Gym: An Evaluation of Temptation Bundling', *Management Science* 60, no. 2 (February 2014): 283–99, https://doi.org/10.1287/mnsc.2013.1784.

6   Brett Tomlinson, 'Behave!' *Princeton Alumni Weekly* (17 October 2016), https://paw.princeton.edu/article/behave-katherine-milkman-04-studies-why-we-do-what-we-do-and-how-change-it.

CHAPTER 21

1   T. C. Sottek, 'Kill the Facebook News Feed', *The Verge* (23 May 2014), https://www.theverge.com/2014/5/23/5744518/kill-the-facebook-news-feed.

2   Freia Lobo, 'This Chrome Extension Makes Your Facebook Addiction Productive', *Mashable* (10 January 2017), http://mashable.com/2017/01/10/todobook-chrome-extension/.

3   https://chrome.google.com/webstore/detail/newsfeed-burner/gdjcjcbjnaelafcijbnceapahcgkpjkl.

4   https://chrome.google.com/webstore/detail/open-multiple-websites/chebdlgebkhbmkeanhkgfojjaofeihgm.

5   Nir Eyal, *Hooked: How to Build Habit-Forming Products*, (New York: Portfolio, 2014).

6   https://chrome.google.com/webstore/detail/df-tube-distraction-free/mjdepdfccjgcndkmemponafgioodelna?hl=en.

CHAPTER 22

1   Lev Grossman, 'Jonathan Franzen: Great American Novelist', *Time* (12 August 2010), http://content.time.com/time/magazine/article/0,9171,2010185-1,00.html.

2   Iain Blair, 'Tarantino Says Horror Movies Are Fun', *Reuters* (5 April 2007), https://www.reuters.com/article/us-tarantino/tarantino-says-horror-movies-are-fun-idUSN2638212720070405.

3   'Booker Prize Nominated Jhumpa Lahiri on India, Being a Mother and Being Inspired by the Ocean', *Harper's Bazaar* (4 October 2013), https://www.harpersbazaar.com/uk/culture/staying-in/news/a20300/booker-prize-nominated-jhumpa-lahiri-on-india-being-a-mother-and-being-inspired-by-the-ocean.

4   Zeb Kurth-Nelson and A. David Redish, 'Don't Let Me Do That! – Models of Precommitment', *Frontiers in Neuroscience* 6, no. 138 (2012), https://doi.org/10.3389/fnins.2012.00138.

5   Adolf Furtwängler, *Odysseus and the Sirens*, detail from an Attic Red-Figured Stamnos, c. 480–470 BC. From Vulci. nd. H. 35.3 cm (13¾ in.), in the British Museum, https://commons.wikimedia.org/wiki/File:Furtwaengler1924009.jpg.

6   'Ulysses Pact', Wikipedia (Accessed 11 February 2017), https://en.wikipedia.org/w/index.php?title=Ulysses_pact&oldid=764886941.

CHAPTER 23

1   https://www.amazon.com/Kitchen-Safe-Locking-Container-Height/dp/B00JGFQTD2.

2   https://selfcontrolapp.com/.

3   https://freedom.to/.

4   https://www.forestapp.cc/.

5   'iOS 12 introduces new features to reduce interruptions and manage Screen Time', Apple Newsroom (4 June 2018), https://www.apple.com/newsroom/2018/06/ios-12-introduces-new-features-to-reduce-interruptions-and-manage-screen-time/.

CHAPTER 24

1   Scott D. Halpern et al., 'Randomized Trial of Four Financial-Incentive Programs for Smoking Cessation', *New England Journal of Medicine* 372, no. 22 (2015): 2108–17, https://doi.org/10.1056/NEJMoa1414293.

CHAPTER 25

1   Christopher J. Bryan et al., 'Motivating Voter Turnout by Invoking the Self', *Proceedings of the National Academy of Sciences* 108, no. 31 (2011): 12653–6, http://dx.doi.org/10.1073/pnas.1103343108.

2   Adam Gorlick, 'Stanford Researchers Find That a Simple Change in Phrasing Can Increase Voter Turnout', *Stanford News* (19 July 2011), http://news.stanford.edu/news/2011/july/increasing-voter-turnout-071911.html.

3   Bryan et al., 'Motivating Voter Turnout by Invoking the Self', op. cit.

4   Vanessa M. Patrick and Henrik Hagtvedt, ' "I Don't" Versus "I Can't": When Empowered Refusal Motivates Goal-Directed Behavior', *Journal of Consumer Research* 39, no. 2 (2012): 371–81, https://doi.org/10.1086/663212.

5   Leah Fessler, 'Psychologists Have Surprising Advice for People Who Feel Unmotivated', *Quartz at Work* (22 August 2018), https://qz.com/work/1363911/two-psychologists-have-a-surprising-theory-on-how-to-get-motivated/.

6   'Targeting Hypocrisy Promotes Safer Sex', *Stanford SPARQ* (Accessed 28 September 2018), https://sparq.stanford.edu/solutions/targeting-hypocrisy-promotes-safer-sex.

7   Lauren Eskreis-Winkler and Ayelet Fishbach, 'Need Motivation at Work? Try Giving Advice', *MIT Sloan Management Review* (13 August 2018), https://sloanreview.mit.edu/article/need-motivation-at-work-try-giving-advice/.

8   Allen Ding Tian et al., 'Enacting Rituals to Improve Self-Control', *Journal of Personality and Social Psychology* 114, no. 6 (2018): 851–76, https://doi.org/10.1037/pspa0000113.

9   Daryl J. Bem, 'Self-Perception Theory', *Advances in Experimental Social Psychology* 6, edited by Leonard Berkowitz, New York: Academic Press, 1972.

10 *The Principles of Psychology*, vol. 2 (New York: Henry Holt and Company, 1918), p. 370.

CHAPTER 26

1 Stephen Stansfeld and Bridget Candy, 'Psychosocial Work Environment and Mental Health: A Meta-Analytic Review', *Scandinavian Journal of Work, Environment & Health* 32, no. 6 (2006): 443–62.

2 Telephone interview with the author, 13 February 2018.

3 'Depression In The Workplace', Mental Health America (1 November 2013), http://www.mentalhealthamerica.net/conditions/depression-workplace.

4 Leslie A. Perlow, *Sleeping with Your Smartphone: How to Break the 24/7 Habit and Change the Way You Work*, Boston, Mass.: Harvard Business Review Press, 2012.

5 Ibid, brackets in the original.

CHAPTER 27

1 Leslie A. Perlow, *Sleeping with Your Smartphone: How to Break the 24/7 Habit and Change the Way You Work*, Boston, Mass.: Harvard Business Review Press, 2012.

2 Julia Rozovsky, 'The Five Keys to a Successful Google Team', *Re:Work* (17 November 2015), https://rework.withgoogle.com/blog/five-keys-to-a-successful-google-team/.

3 Amy Edmondson, 'Building a Psychologically Safe Workplace', TEDx talk at TEDxHGSE (4 May 2014), https://www.youtube.com/watch?time_continue=231&v=LhoLuui9gX8.

4 Ibid.

CHAPTER 28

1 'With 10+ Million Daily Active Users, Slack Is Where More Work Happens Every Day, All Over the World', The Official Slack Blog (Accessed 22 March 2019), https://slackhq.com/slack-has-10-million-daily-active-users.

2 Jeff Bercovici, 'Slack Is Our Company of the Year. Here's Why Everybody's Talking About It', *Inc.* (23 November 2015), https://www.inc.com/magazine/201512/jeff-bercovici/slack-company-of-the-year-2015.html.

3 Casey Renner, 'Former Slack CMO, Bill Macaitis, on How Slack Uses Slack', *OpenView Labs* (19 May 2017), https://labs.openviewpartners.com/how-slack-uses-slack/.

4 Graeme Codrington, 'Good to Great ... to Gone!', *Tomorrow Trends* (9 December 2011), http://www.tomorrowtodayglobal.com/2011/12/09/good-to-great-to-gone-2/.

5 Boston Consulting Group Overview on Glassdoor (Accessed 12 February 2018), https://www.glassdoor.com/Overview/Working-at-Boston-Consulting-Group-EI_IE3879.11,34.htm.

6 Slack Reviews on Glassdoor (Accessed 12 February 2018), https://www.glassdoor.com/Reviews/slack-reviews-SRCH_KE0,5.htm.

CHAPTER 29

1 Jean M. Twenge, 'Have Smartphones Destroyed a Generation?' *The Atlantic* (September 2017), https://www.theatlantic.com/magazine/archive/2017/09/has-the-smartphone-destroyed-a-generation/534198/.

2 'The Risk Of Teen Depression And Suicide Is Linked To Smartphone Use, Study Says', NPR.org (Accessed 6 January 2019), https://www.npr.org/2017/12/17/571443683/the-call-in-teens-and-depression.

3 Twenge, op. cit.

4 YouTube search, 'dad destroys kids phone', YouTube (Accessed 23 July 2018), https://www.youtube.com/results?search_query=dad+destroys+kids+phone.

5 Mark L. Wolraich, David B. Wilson and J. Wade White, 'The Effect of Sugar on Behavior or Cognition in Children: A Meta-Analysis', *JAMA* 274, no. 20 (22 November 1995): 1617–21, https://doi.org/10.1001/jama.1995.03530200053037.

6 Alice Schlegel and Herbert Barry III, 'Adolescence: An Anthropological Inquiry', *Journal of Nervous and Mental Disease* 180 (1 May 1992), https://doi.org/10.2307/2076290.

7 Robert Epstein, 'The Myth of the Teen Brain', *Scientific American* (1 June 2007), https://www.scientificamerican.com/article/the-myth-of-the-teen-brain-2007-06/.

8 Richard McSherry, 'Suicide and Homicide Under Insidious Forms', *Sanitarian*, 26 April 1883.

9   W. W. J., review of *Children and Radio Programs: A Study of More than Three Thousand Children in the New York Metropolitan Area*, by Azriel L. Eisenberg, *Gramophone*, September 1936, https://reader.exacteditions.com/issues/32669/page/31?term=crime.

10   Abigail Wills, 'Youth Culture and Crime: What Can We Learn from History?' *History Extra*, 12 August 2009, www.historyextra.com/period/20th-century/youth-culture- and-crime-what-can-we-learn-from-history/.

11   'No, Smartphones Are Not Destroying a Generation', *Psychology Today* (Accessed 7 January 2019), https://www.psychologytoday.com/blog/once-more-feeling/201708/no-smartphones-are-not-destroying-generation.

12   'More Screen Time for Kids Isn't All That Bad: Researcher Says Children Should Be Allowed to Delve into Screen Technology, as It Is Becoming an Essential Part of Modern Life', *Science Daily* (Accessed 7 January 2019), https://www.sciencedaily.com/releases/2017/02/170207105326.htm.

13   'A Large-Scale Test of the Goldilocks Hypothesis: Quantifying the Relations Between Digital-Screen Use and the Mental Well-Being of Adolescents – Andrew K. Przybylski, Netta Weinstein, 2017', *Psychological Science* 28, no. 2, 13 January 2017 (Accessed 7 January 2019), https://journals.sagepub.com/doi/10.1177/0956797616678438.

14   'It Turns Out Staring At Screens Isn't Bad For Teens' Mental Wellbeing' *Buzzfeed*, 14 January 2017 (Accessed 7 January 2019), https://www.buzzfeed.com/tomchivers/mario-kart-should-be-available-on-the-nhs.

CHAPTER 30

1   R. M. Ryan and E. L. Deci, 'Self-Determination Theory and the Facilitation of Intrinsic Motivation, Social Development, and Well-Being', *American Psychologist* 55 (2000): 68–78. https://dx.doi.org/10.1037/0003-066X.55.1.68.

2   Maricela Correa-Chávez and Barbara Rogoff, 'Children's Attention to Interactions Directed to Others: Guatemalan Mayan and

European American Patterns', *Developmental Psychology* 45, no. 3 (May 2009): 630–41, https://doi.org/10.1037/a0014144.

3 Michaeleen Doucleff, 'A Lost Secret: How To Get Kids To Pay Attention', *NPR* (21 June 2018), https://www.npr.org/sections/goatsandsoda/2018/06/21/621752789/a-lost-secret-how-to-get-kids-to-pay-attention.

4 'A Lost Secret: How To Get Kids To Pay Attention', *NPR*.

5 Research assistant interview with Richard Ryan, May 2017.

6 Robert Epstein, 'The Myth of the Teen Brain', *Scientific American* (1 June 2007), https://www.scientificamerican.com/article/the-myth-of-the-teen-brain-2007-06/.

7 Interview with Ryan, May 2017.

8 Peter Gray, 'The Decline of Play and the Rise of Psychopathy in Children and Adolescents', *American Journal of Play* 3, no. 4 (Spring 2011): pp. 443–63.

9 Esther Entin, 'All Work and No Play: Why Your Kids Are More Anxious, Depressed', *The Atlantic* (12 October 2011), https://www.theatlantic.com/health/archive/2011/10/all-work-and-no-play-why-your-kids-are-more-anxious-depressed/246422/.

10 'There's Never Been a Safer Time to Be a Kid in America', *Washington Post* (Accessed 23 March 2019), https://www.washingtonpost.com/news/wonk/wp/2015/04/14/theres-never-been-a-safer-time-to-be-a-kid-in-america/.

11 Research assistant interview with Richard Ryan, May 2017.

12 Gray, op. cit.

13 Interview with Ryan, May 2017.

14 Richard M. Ryan and Edward L. Deci, *Self-Determination Theory: Basic Psychological Needs in Motivation, Development, and Wellness*, New York: Guilford Publications, 2017, p. 524.

CHAPTER 31

1 Research assistant interview with Lori Getz and family, May 2017.

2 Alison Gopnik, 'Playing Is More Than Fun – It's Smart', *The Atlantic* (12 August 2016), https://www.theatlantic.com/education/archive/2016/08/in-defense-of-play/495545/.

3   Anne Fishel, 'The Most Important Thing You Can Do with Your Kids? Eat Dinner with Them', *Washington Post* (12 January 2015), https://www.washingtonpost.com/posteverything/wp/2015/01/12/the-most-important-thing-you-can-do-with-your-kids-eat-dinner-with-them/.

CHAPTER 32

1   'Teens, Social Media & Technology 2018' Pew Research Center, 31 May 2018, http://www.pewinternet.org/2018/05/31/teens-social-media-technology-2018/.

2   'Mobile Kids: The Parent, the Child and the Smartphone' (Accessed 12 January 2019), https://www.nielsen.com/us/en/insights/news/2017/mobile-kids--the-parent-the-child-and-the-smartphone.html.

3   AIEK/AEKU X8 Ultra Thin Card Mobile Phone Mini Pocket Students Phone Low Radiation Support TF Card PK AIEK E1 X6 M5 C6-in Mobile Phones from Cellphones & Telecommunications on Aliexpress.Com | Alibaba Group.' aliexpress.com. (Accessed 12 January 2019), https://www.aliexpress.com/item/New-AIEK-AEKU-X8-Ultra-Thin-Card-Mobile-Phone-Mini-Pocket-Students-Phone-Low-Radiation-Support/32799743043.html.

4   'Verizon's $180 GizmoWatch Lets Parents Track Kids' Location and Activity', *CNET* (Accessed 12 January 2019), https://www.cnet.com/news/verizons-180-gizmowatch-lets-parents-track-kids-location-activity/.

5   Anya Kamenetz, *The Art of Screen Time: How Your Family Can Balance Digital Media and Real Life*, New York: PublicAffairs, 2018.

CHAPTER 34

1   Nicholas A. Christakis and James H. Fowler, 'Social Contagion Theory: Examining Dynamic Social Networks and Human Behavior', *Statistics in Medicine* 32, no. 4 (20 February 2013), 556–77, https://doi.org/10.1002/sim.5408.

2   Kelly Servick, 'Should We Treat Obesity like a Contagious Disease?' *Science* (19 February 2017), http://www.sciencemag.org/news/2017/02/should-we-treat-obesity-contagious-disease.

3   Paul Graham, 'The Acceleration of Addictiveness' (Accessed 6 December 2017), http://www.paulgraham.com/addiction.html.6.

4   'Trends in Current Cigarette Smoking Among High School Students and Adults, United States, 1965–2014', Centers for Disease Control and Prevention (Accessed 6 December 2017), http://www.cdc.gov/tobacco/data_statistics/tables/trends/cig_smoking/.

5   'Macquarie, "Phubbing: A Word Is Born" // McCann Melbourne' (26 June 2014), https://www.youtube.com/watch?v=hLNhKUniaEw.

CHAPTER 35

1   Rich Miller, 'Give Up Sex or Your Mobile Phone? Third of Americans Forgo Sex', *Bloomberg* (15 January 2015), https://www.bloomberg.com/news/articles/2015-01-15/give-up-sex-or-your-mobile-phone-third-of-americans-forgo-sex.

2   Russell Heimlich, 'Do You Sleep With Your Cell Phone?' Pew Research Center (blog) (Accessed 15 January 2019), http://www.pewresearch.org/fact-tank/2010/09/13/do-you-sleep-with-your-cell-phone/.

3   https://eero.com.

4   *New Oxford American Dictionary* (2nd edn).

A NOTE ON THE AUTHORS

Nir Eyal taught behavioural design at the Stanford Graduate School of Business and the Hasso Plattner Institute of Design at Stanford. He writes, consults and teaches about the intersection of psychology, technology and business at NirAndFar.com. His writing has been featured in *Harvard Business Review, The Atlantic, Time, The Week, Inc.* and *Psychology Today.*

His 2014 book *Hooked: How to Build Habit-Forming Products* is a *Wall Street Journal* bestseller, has been translated into over eighteen languages and won the 'Marketing Book of the Year' award from 800 CEO Read.

Julie Li co-founded NirAndFar.com, where she works to bring the latest insights on time management, behavioural design and consumer psychology to a growing global audience. Julie previously co-founded two start-ups and helped lead both companies to acquisition.

Nir Eyal is the internationally bestselling author of *Hooked*: *How to Build Habit-Forming Products*. He has taught courses on applied consumer psychology at the Stanford Graduate School of Business and the Hasso Plattner Institute of Design, and is a frequent speaker at industry conferences and Fortune 500 companies. His writing on technology, psychology and business appears in the *Harvard Business Review*, the *Atlantic*, *TechCrunch* and *Psychology Today*.

To get in touch with Nir, visit nirandfar.com

We are living through a crisis of distraction. Plans get sidetracked, friends are ignored, work never seems to get done.

Why does it feel like we're distracting our lives away?

In *Indistractable*, behavioural designer Nir Eyal shows what life could look like if you followed through on your intentions. Instead of suggesting a digital detox, Eyal reveals the hidden psychology driving you to distraction, and teaches you how to make pacts with yourself to keep your brain on track. *Indistractable* is a guide to making decisions and seeing them through.

Empowering and optimistic, this is the book that will help you design your time, realise your ambitions, and live the life you really want.

INSTRUCTIONS

1
Cut the cards along the solid lines.

2
Cut the right slot depending on the thickness of your monitor and adjust it as you see fit.

3
Fold the cards along the dotted lines.

4
Glue or tape it together.

5
Place it on your monitor when you need to focus.

6
Visit INDISTRACTABLE.COM for more.

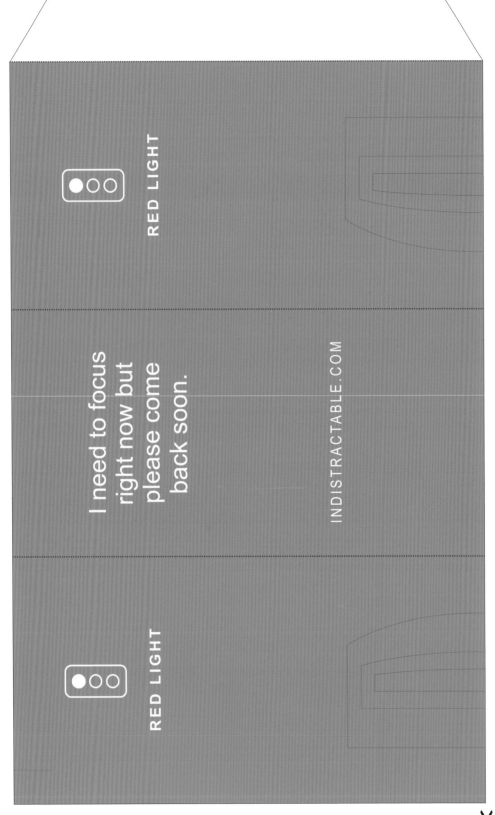

RED LIGHT

I need to focus
right now but
please come
back soon.

INDISTRACTABLE.COM

RED LIGHT